Fundamentals of
Computational Chemistry

Fundamentals of
Computational Chemistry

Editor: Daniel Sullivan

NYRESEARCH
P R E S S

New York

Published by NY Research Press
118-35 Queens Blvd., Suite 400,
Forest Hills, NY 11375, USA
www.nyresearchpress.com

Fundamentals of Computational Chemistry
Edited by Daniel Sullivan

International Standard Book Number: 978-1-63238-585-7 (Hardback)

Cataloging-in-Publication Data

Fundamentals of computational chemistry / edited by Daniel Sullivan.
    p. cm.
Includes bibliographical references and index.
ISBN 978-1-63238-585-7
1. Chemistry--Data processing. 2. Chemistry--Computer simulation.
3. Chemistry--Mathematics. I. Sullivan, Daniel.
QD39.3.E46 F86 2018
542.85--dc23

Contents

Permissions

Index

Preface

Computational chemistry is a sub-field of chemistry which aims at solving problems of chemical nature by using computer programs. It calculates molecular structure and properties by integrating methods of theoretical chemistry and computer programs. Some of the methods used in this field are ab initio methods, chemical and molecular dynamics, density functional methods, etc. This book elucidates the concepts and innovative models around prospective developments with respect to computational chemistry. It is a compilation of chapters that discuss the most vital concepts in this field. For all those who are interested in this subject, this textbook can prove to be an essential guide.

To facilitate a deeper understanding of the contents of this book a short introduction of every chapter is written below:

Chapter 1- Computational chemistry uses mathematical algorithms and statistics for chemical problems. Computational methods can predict complex and unsolved chemical problems. This field is used in pharmaceutics and drug design. The chapter on computational chemistry offers an insightful focus, keeping in mind the complex subject matter.

Chapter 2- The methods used in computational chemistry are ab initio quantum chemistry methods, quantum chemistry composite methods, Møller–Plesset perturbation theory etc. Ab initio quantum chemistry method is entirely based on quantum chemistry. The aspects elucidated in this section are of vital importance, and provide a better understanding of ab initio methods.

Chapter 3- Molecular dynamics helps in developing a better understanding of the movement of atoms and molecules. Computer simulation is used in this process. Tools and techniques are an important component of any field of study. The following chapter elucidates the various tools and techniques that are related to molecular dynamics.

Chapter 4- Force fields in molecular modeling refer to the functional form and parameter sets that are established in order to compute the potential energy of a system of atoms. Classical force fields include Assisted Model Building and Energy Refinement (AMBER), Chemistry at HARvard Molecular Mechanics (CHARMM), etc. Computational chemistry is best understood in confluence with the major topics listed in the following chapter.

Chapter 5- Some of the software used in computational chemistry include the Advanced Simulation Library (ASL), Crystal, Gaussian, etc. Gaussian can perform functions pertaining to AMBER, Universal Force Field (UFF), Hartree-Fock method among others. The topics discussed in the chapter are of great importance to broaden the existing knowledge on computational chemistry.

I would like to share the credit of this book with my editorial team who worked tirelessly on this book. I owe the completion of this book to the never-ending support of my family, who supported me throughout the project.

Editor

A Brief Introduction to Computational Chemistry

Computational chemistry uses mathematical algorithms and statistics for chemical problems. Computational methods can predict complex and unsolved chemical problems. This field is used in pharmaceutics and drug design. The chapter on computational chemistry offers an insightful focus, keeping in mind the complex subject matter.

Computational Chemistry

Computational chemistry is a branch of chemistry that uses computer simulation to assist in solving chemical problems. It uses methods of theoretical chemistry, incorporated into efficient computer programs, to calculate the structures and properties of molecules and solids. It is necessary because, apart from relatively recent results concerning the hydrogen molecular ion (dihydrogen cation), the quantum many-body problem cannot be solved analytically, much less in closed form. While computational results normally complement the information obtained by chemical experiments, it can in some cases predict hitherto unobserved chemical phenomena. It is widely used in the design of new drugs and materials.

Examples of such properties are structure (i.e., the expected positions of the constituent atoms), absolute and relative (interaction) energies, electronic charge density distributions, dipoles and higher multipole moments, vibrational frequencies, reactivity, or other spectroscopic quantities, and cross sections for collision with other particles.

The methods used cover both static and dynamic situations. In all cases, the computer time and other resources (such as memory and disk space) increase rapidly with the size of the system being studied. That system can be one molecule, a group of molecules, or a solid. Computational chemistry methods range from very approximate to highly accurate; the latter are usually feasible for small systems only. *Ab initio* methods are based entirely on quantum mechanics and basic physical constants. Other methods are called empirical or semi-empirical because they use additional empirical parameters.

Both *ab initio* and semi-empirical approaches involve approximations. These range from simplified forms of the first-principles equations that are easier or faster to solve, to approximations limiting the size of the system (for example, periodic boundary conditions), to fundamental approximations to the underlying equations that are required to achieve any solution to them at all. For example, most *ab initio* calculations make the Born–Oppenheimer approximation, which greatly simplifies the underlying Schrödinger equation by assuming that the nuclei remain in place during the calculation. In principle, *ab initio* methods eventually converge to the exact solution of the underlying equations as the number of approximations is reduced. In practice, however, it is impossible to eliminate all approximations, and residual error inevitably remains. The goal of computational chemistry is to minimize this residual error while keeping the calculations tractable.

In some cases, the details of electronic structure are less important than the long-time phase space behavior of molecules. This is the case in conformational studies of proteins and protein-ligand binding thermodynamics. Classical approximations to the potential energy surface are used, as they are computationally less intensive than electronic calculations, to enable longer simulations of molecular dynamics. Furthermore, cheminformatics uses even more empirical (and computationally cheaper) methods like machine learning based on physicochemical properties. One typical problem in cheminformatics is to predict the binding affinity of drug molecules to a given target.

History

Building on the founding discoveries and theories in the history of quantum mechanics, the first theoretical calculations in chemistry were those of Walter Heitler and Fritz London in 1927. The books that were influential in the early development of computational quantum chemistry include Linus Pauling and E. Bright Wilson's 1935 *Introduction to Quantum Mechanics – with Applications to Chemistry*, Eyring, Walter and Kimball's 1944 *Quantum Chemistry*, Heitler's 1945 *Elementary Wave Mechanics – with Applications to Quantum Chemistry*, and later Coulson's 1952 textbook *Valence*, each of which served as primary references for chemists in the decades to follow.

With the development of efficient computer technology in the 1940s, the solutions of elaborate wave equations for complex atomic systems began to be a realizable objective. In the early 1950s, the first semi-empirical atomic orbital calculations were performed. Theoretical chemists became extensive users of the early digital computers. A very detailed account of such use in the United Kingdom is given by Smith and Sutcliffe. The first *ab initio* Hartree–Fock method calculations on diatomic molecules were performed in 1956 at MIT, using a basis set of Slater orbitals. For diatomic molecules, a systematic study using a minimum basis set and the first calculation with a larger basis set were published by Ransil and Nesbet respectively in 1960. The first polyatomic calculations using Gaussian orbitals were performed in the late 1950s. The first configuration interaction calculations were performed in Cambridge on the EDSAC computer in the 1950s using Gaussian orbitals by Boys and coworkers. By 1971, when a bibliography of *ab initio* calculations was published, the largest molecules included were naphthalene and azulene. Abstracts of many earlier developments in *ab initio* theory have been published by Schaefer.

In 1964, Hückel method calculations (using a simple linear combination of atomic orbitals (LCAO) method to determine electron energies of molecular orbitals of π electrons in conjugated hydrocarbon systems) of molecules, ranging in complexity from butadiene and benzene to ovalene, were generated on computers at Berkeley and Oxford. These empirical methods were replaced in the 1960s by semi-empirical methods such as CNDO.

In the early 1970s, efficient *ab initio* computer programs such as ATMOL, Gaussian, IBMOL, and POLYAYTOM, began to be used to speed *ab initio* calculations of molecular orbitals. Of these four programs, only Gaussian, now vastly expanded, is still in use, but many other programs are now in use. At the same time, the methods of molecular mechanics, such as MM2 force field, were developed, primarily by Norman Allinger.

One of the first mentions of the term *computational chemistry* can be found in the 1970 book *Computers and Their Role in the Physical Sciences* by Sidney Fernbach and Abraham Haskell Taub, where they state "It seems, therefore, that 'computational chemistry' can finally be more and more of a reality." During the 1970s, widely different methods began to be seen as part of a new emerging discipline of *computational chemistry*. The *Journal of Computational Chemistry* was first published in 1980.

Computational chemistry has featured in several Nobel Prize awards, most notably in 1998 and 2013. Walter Kohn, "for his development of the density-functional theory", and John Pople, "for his development of computational methods in quantum chemistry", received the 1998 Nobel Prize in Chemistry. Martin Karplus, Michael Levitt and Arieh Warshel received the 2013 Nobel Prize in Chemistry for "the development of multiscale models for complex chemical systems".

Fields of Application

The term *theoretical chemistry* may be defined as a mathematical description of chemistry, whereas *computational chemistry* is usually used when a mathematical method is sufficiently well developed that it can be automated for implementation on a computer. In theoretical chemistry, chemists, physicists, and mathematicians develop algorithms and computer programs to predict atomic and molecular properties and reaction paths for chemical reactions. Computational chemists, in contrast, may simply apply existing computer programs and methodologies to specific chemical questions.

Computational chemistry has two different aspects:

- Computational studies, used to find a starting point for a laboratory synthesis, or to assist in understanding experimental data, such as the position and source of spectroscopic peaks.

- Computational studies, used to predict the possibility of so far entirely unknown molecules or to explore reaction mechanisms not readily studied via experiments.

Thus, computational chemistry can assist the experimental chemist or it can challenge the experimental chemist to find entirely new chemical objects.

Several major areas may be distinguished within computational chemistry:

- The prediction of the molecular structure of molecules by the use of the simulation of forces, or more accurate quantum chemical methods, to find stationary points on the energy surface as the position of the nuclei is varied.

- Storing and searching for data on chemical entities.

- Identifying correlations between chemical structures and properties.

- Computational approaches to help in the efficient synthesis of compounds.

- Computational approaches to design molecules that interact in specific ways with other molecules (e.g. drug design and catalysis).

Accuracy

The words *exact* and *perfect* do not apply here, as very few aspects of chemistry can be computed exactly. However, almost every aspect of chemistry can be described in a qualitative or approximate quantitative computational scheme.

Molecules consist of nuclei and electrons, so the methods of quantum mechanics apply. Computational chemists often attempt to solve the non-relativistic Schrödinger equation, with relativistic corrections added, although some progress has been made in solving the fully relativistic Dirac equation. In principle, it is possible to solve the Schrödinger equation in either its time-dependent or time-independent form, as appropriate for the problem in hand; in practice, this is not possible except for very small systems. Therefore, a great number of approximate methods strive to achieve the best trade-off between accuracy and computational cost.

Accuracy can always be improved with greater computational cost. Significant errors can present themselves in ab initio models comprising many electrons, due to the computational cost of full relativistic-inclusive methods. This complicates the study of molecules interacting with high atomic mass unit atoms, such as transitional metals and their catalytic properties. Present algorithms in computational chemistry can routinely calculate the properties of molecules that contain up to about 40 electrons with sufficient accuracy. Errors for energies can be less than a few kJ/mol. For geometries, bond lengths can be predicted within a few picometres and bond angles within 0.5 degrees. The treatment of larger molecules that contain a few dozen electrons is computationally tractable by approximate methods such as density functional theory (DFT).

There is some dispute within the field whether or not the latter methods are sufficient to describe complex chemical reactions, such as those in biochemistry. Large molecules can be studied by semi-empirical approximate methods. Even larger molecules are treated by classical mechanics methods that use what are called molecular mechanics (MM). In QM-MM methods, small parts of large complexes are treated quantum mechanically (QM), and the remainder is treated approximately (MM).

Methods

One molecular formula can represent more than one molecular isomer: a set of isomers. Each isomer is a local minimum on the energy surface (called the potential energy surface) created from the total energy (i.e., the electronic energy, plus the repulsion energy between the nuclei) as a function of the coordinates of all the nuclei. A stationary point is a geometry such that the derivative of the energy with respect to all displacements of the nuclei is zero. A local (energy) minimum is a stationary point where all such displacements lead to an increase in energy. The local minimum that is lowest is called the global minimum and corresponds to the most stable isomer. If there is one particular coordinate change that leads to a decrease in the total energy in both directions, the stationary point is a transition structure and the coordinate is the reaction coordinate. This process of determining stationary points is called geometry optimization.

The determination of molecular structure by geometry optimization became routine only after efficient methods for calculating the first derivatives of the energy with respect to all atomic coordinates became available. Evaluation of the related second derivatives allows the prediction of vibrational frequencies if harmonic motion is estimated. More importantly, it allows for the characterization of stationary points. The frequencies are related to the eigenvalues of the Hessian matrix, which contains second derivatives. If the eigenvalues are all positive, then the frequencies are all real and the stationary point is a local minimum. If one eigenvalue is negative (i.e., an imaginary frequency), then the stationary point is a transition structure. If more than one eigenvalue is negative, then the stationary point is a more complex one, and is usually of little interest. When one of these is found, it is necessary to move the search away from it if the experimenter is looking solely for local minima and transition structures.

The total energy is determined by approximate solutions of the time-dependent Schrödinger equation, usually with no relativistic terms included, and by making use of the Born–Oppenheimer approximation, which allows for the separation of electronic and nuclear motions, thereby simplifying the Schrödinger equation. This leads to the evaluation of the total energy as a sum of the electronic energy at fixed nuclei positions and the repulsion energy of the nuclei. A notable exception are certain approaches called direct quantum chemistry, which treat electrons and nuclei on a common footing. Density functional methods and semi-empirical methods are variants on the major theme. For very large systems, the relative total energies can be compared using molecular mechanics. The ways of determining the total energy to predict molecular structures are:

Ab Initio Methods

The programs used in computational chemistry are based on many different quan-tum-chemical methods that solve the molecular Schrödinger equation associated with the molecular Hamiltonian. Methods that do not include any empirical or semi-empirical parameters in their equations – being derived directly from theoretical principles, with no inclusion of experimental data – are called *ab initio methods*. This does not imply that the solution is an exact one; they are all approximate quantum mechanical calculations. It means that a particular approximation is rigorously defined on first principles (quan-tum theory) and then solved within an error margin that is qualitatively known before-hand. If numerical iterative methods must be used, the aim is to iterate until full machine accuracy is obtained (the best that is possible with a finite word length on the computer, and within the mathematical and/or physical approximations made).

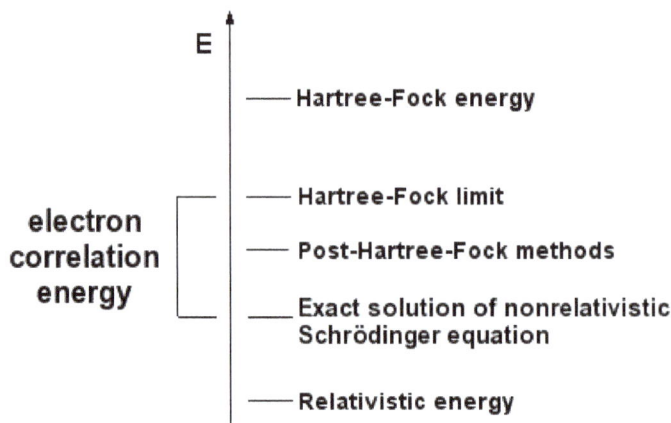

Diagram illustrating various *ab initio* electronic structure methods in terms of energy.
Spacings are not to scale.

The simplest type of *ab initio* electronic structure calculation is the Hartree–Fock method (HF), an extension of molecular orbital theory, in which the correlated elec-tron-electron repulsion is not specifically taken into account; only its average effect is included in the calculation. As the basis set size is increased, the energy and wave function tend towards a limit called the Hartree–Fock limit. Many types of calcula-tions (termed post-Hartree–Fock methods) begin with a Hartree–Fock calculation and subsequently correct for electron-electron repulsion, referred to also as electron-ic correlation. As these methods are pushed to the limit, they approach the exact solution of the non-relativistic Schrödinger equation. To obtain exact agreement with experiment, it is necessary to include relativistic and spin orbit terms, both of which are far more important for heavy atoms. In all of these approaches, along with choice of method, it is necessary to choose a basis set. This is a set of functions, usually cen-tered on the different atoms in the molecule, which are used to expand the molecu-lar orbitals with the linear combination of atomic orbitals (LCAO) molecular orbital method ansatz. Ab initio methods need to define a level of theory (the method) and a basis set.

The Hartree–Fock wave function is a single configuration or determinant. In some cases, particularly for bond breaking processes, this is inadequate, and several configurations must be used. Here, the coefficients of the configurations, and of the basis functions, are optimized together.

The total molecular energy can be evaluated as a function of the molecular geometry; in other words, the potential energy surface. Such a surface can be used for reaction dynamics. The stationary points of the surface lead to predictions of different isomers and the transition structures for conversion between isomers, but these can be determined without a full knowledge of the complete surface.

A particularly important objective, called computational thermochemistry, is to calculate thermochemical quantities such as the enthalpy of formation to chemical accuracy. Chemical accuracy is the accuracy required to make realistic chemical predictions and is generally considered to be 1 kcal/mol or 4 kJ/mol. To reach that accuracy in an economic way it is necessary to use a series of post-Hartree–Fock methods and combine the results. These methods are called quantum chemistry composite methods.

Density Functional Methods

Density functional theory (DFT) methods are often considered to be *ab initio methods* for determining the molecular electronic structure, even though many of the most common functionals use parameters derived from empirical data, or from more complex calculations. In DFT, the total energy is expressed in terms of the total one-electron density rather than the wave function. In this type of calculation, there is an approximate Hamiltonian and an approximate expression for the total electron density. DFT methods can be very accurate for little computational cost. Some methods combine the density functional exchange functional with the Hartree–Fock exchange term and are termed hybrid functional methods.

Semi-empirical and Empirical Methods

Semi-empirical quantum chemistry methods are based on the Hartree–Fock method formalism, but make many approximations and obtain some parameters from empirical data. They are very important in computational chemistry for treating large molecules where the full Hartree–Fock method without the approximations is too costly. The use of empirical parameters appears to allow some inclusion of correlation effects into the methods.

Semi-empirical methods follow what are often called empirical methods, where the two-electron part of the Hamiltonian is not explicitly included. For π-electron systems, this was the Hückel method proposed by Erich Hückel, and for all valence electron systems, the extended Hückel method proposed by Roald Hoffmann.

Molecular Mechanics

In many cases, large molecular systems can be modeled successfully while avoiding quantum mechanical calculations entirely. Molecular mechanics simulations, for example, use one classical expression for the energy of a compound, for instance the harmonic oscillator. All constants appearing in the equations must be obtained beforehand from experimental data or *ab initio* calculations.

The database of compounds used for parameterization, i.e., the resulting set of parameters and functions is called the force field, is crucial to the success of molecular mechanics calculations. A force field parameterized against a specific class of molecules, for instance proteins, would be expected to only have any relevance when describing other molecules of the same class.

These methods can be applied to proteins and other large biological molecules, and allow studies of the approach and interaction (docking) of potential drug molecules.

Methods for Solids

Computational chemical methods can be applied to solid state physics problems. The electronic structure of a crystal is in general described by a band structure, which defines the energies of electron orbitals for each point in the Brillouin zone. Ab initio and semi-empirical calculations yield orbital energies; therefore, they can be applied to band structure calculations. Since it is time-consuming to calculate the energy for a molecule, it is even more time-consuming to calculate them for the entire list of points in the Brillouin zone.

Chemical Dynamics

Once the electronic and nuclear variables are separated (within the Born–Oppenheimer representation), in the time-dependent approach, the wave packet corresponding to the nuclear degrees of freedom is propagated via the time evolution operator (physics) associated to the time-dependent Schrödinger equation (for the full molecular Hamiltonian). In the complementary energy-dependent approach, the time-independent Schrödinger equation is solved using the scattering theory formalism. The potential representing the interatomic interaction is given by the potential energy surfaces. In general, the potential energy surfaces are coupled via the vibronic coupling terms.

The most popular methods for propagating the wave packet associated to the molecular geometry are:

- the split operator technique,

- the Chebyshev (real) polynomial,

- the multi-configuration time-dependent Hartree method (MCTDH),

- the semiclassical method.

Molecular Dynamics

Molecular dynamics (MD) use either quantum mechanics, Newton's laws of motion or a mixed model to examine the time-dependent behavior of systems, including vibrations or Brownian motion and reactions. MD combined with density functional theory leads to hybrid models.

Interpreting Molecular Wave Functions

The atoms in molecules (QTAIM) model of Richard Bader was developed to effectively link the quantum mechanical model of a molecule, as an electronic wavefunction, to chemically useful concepts such as atoms in molecules, functional groups, bonding, the theory of Lewis pairs, and the valence bond model. Bader has demonstrated that these empirically useful chemistry concepts can be related to the topology of the observable charge density distribution, whether measured or calculated from a quantum mechanical wavefunction. QTAIM analysis of molecular wavefunctions is implemented, for example, in the AIMAll software package.

Software Packages

Many self-sufficient computational chemistry software packages exist. Some include many methods covering a wide range, while others concentrate on a very specific range or even on one method. Details of most of them can be found in:

- Biomolecular modelling programs: proteins, nucleic acid.

- Molecular mechanics programs.

- Quantum chemistry and solid state physics software supporting several methods.

- Molecular design software

- Semi-empirical programs.

- Valence bond programs.

Cheminformatics

Cheminformatics (also known as chemoinformatics, chemioinformatics and chemical informatics) is the use of computer and informational techniques applied to a range of problems in the field of chemistry. These *in silico* techniques are used, for example, in pharmaceutical companies in the process of drug discovery. These methods can also be used in chemical and allied industries in various other forms.

History

The term chemoinformatics was defined by F.K. Brown in 1998:

Chemoinformatics is the mixing of those information resources to transform data into information and information into knowledge for the intended purpose of making better decisions faster in the area of drug lead identification and optimization.

Since then, both spellings have been used, and some have evolved to be established as Cheminformatics, while European Academia settled in 2006 for Chemoinformatics. The recent establishment of the Journal of Cheminformatics is a strong push towards the shorter variant.

Basics

Cheminformatics combines the scientific working fields of chemistry, computer science and information science for example in the areas of topology, chemical graph theory, information retrieval and data mining in the chemical space. Cheminformatics can also be applied to data analysis for various industries like paper and pulp, dyes and such allied industries.

Applications

Storage and Retrieval

The primary application of cheminformatics is in the storage, indexing and search of information relating to compounds. The efficient search of such stored information includes topics that are dealt with in computer science as data mining, information retrieval, information extraction and machine learning. Related research topics include:

- Unstructured data
  - Information retrieval
  - Information extraction
- Structured data mining and mining of structured data
  - Database mining
  - Graph mining
  - Molecule mining
  - Sequence mining
  - Tree mining
- Digital libraries

File Formats

The *in silico* representation of chemical structures uses specialized formats such as the XML-based Chemical Markup Language or SMILES. These representations are often used for storage in large chemical databases. While some formats are suited for visual representations in 2 or 3 dimensions, others are more suited for studying physical interactions, modeling and docking studies.

Virtual Libraries

Chemical data can pertain to real or virtual molecules. Virtual libraries of compounds may be generated in various ways to explore chemical space and hypothesize novel compounds with desired properties.

Virtual libraries of classes of compounds (drugs, natural products, diversity-oriented synthetic products) were recently generated using the FOG (fragment optimized growth) algorithm. This was done by using cheminformatic tools to train transition probabilities of a Markov chain on authentic classes of compounds, and then using the Markov chain to generate novel compounds that were similar to the training database.

Virtual Screening

In contrast to high-throughput screening, virtual screening involves computationally screening *in silico* libraries of compounds, by means of various methods such as docking, to identify members likely to possess desired properties such as biological activity against a given target. In some cases, combinatorial chemistry is used in the development of the library to increase the efficiency in mining the chemical space. More commonly, a diverse library of small molecules or natural products is screened.

Quantitative Structure-activity Relationship (QSAR)

This is the calculation of quantitative structure-activity relationship and quantitative structure property relationship values, used to predict the activity of compounds from their structures. In this context there is also a strong relationship to chemometrics. Chemical expert systems are also relevant, since they represent parts of chemical knowledge as an *in silico* representation. There is a relatively new concept of matched molecular pair analysis or prediction-driven MMPA which is coupled with QSAR model in order to identify activity cliff.

Chemical Database

A chemical database is a database specifically designed to store chemical information. This information is about chemical and crystal structures, spectra, reactions and syntheses, and thermophysical data.

Types of Chemical Databases

Chemical Structures

Chemical structures are traditionally represented using lines indicating chemical bonds between atoms and drawn on paper (2D structural formulae). While these are ideal visual representations for the chemist, they are unsuitable for computational use and especially for search and storage. Small molecules (also called ligands in drug design applications), are usually represented using lists of atoms and their connections. Large molecules such as proteins are however more compactly represented using the sequences of their amino acid building blocks. Large chemical databases for structures are expected to handle the storage and searching of information on millions of molecules taking terabytes of physical memory.

Literature Database

Chemical literature databases correlate structures or other chemical information to relevant references such as academic papers or patents. This type of database includes STN, Scifinder, and Reaxys. Links to literature are also included in many databases that focus on chemical characterization.

Crystallographic Database

Crystallographic databases store X-ray crystal structure data. Common examples include Protein Data Bank and Cambridge Structural Database.

NMR Spectra Database

NMR spectra databases correlate chemical structure with NMR data. These databases often include other characterization data such as FTIR and mass spectrometry.

Reactions Database

Most chemical databases store information on stable molecules but in databases for reactions also intermediates and temporarily created unstable molecules are stored. Reaction databases contain information about products, educts, and reaction mechanisms.

Thermophysical Database

Thermophysical data are information about

- phase equilibria including vapor–liquid equilibrium, solubility of gases in liquids, liquids in solids (SLE), heats of mixing, vaporization, and fusion.
- caloric data like heat capacity, heat of formation and combustion,
- transport properties like viscosity and thermal conductivity

Chemical Structure Representation

There are two principal techniques for representing chemical structures in digital databases

- As connection tables / adjacency matrices / lists with additional information on bond (edges) and atom attributes (nodes), such as:

  MDL Molfile, PDB, CML

- As a linear string notation based on depth first or breadth first traversal, such as:

  SMILES/SMARTS, SLN, WLN, InChI

These approaches have been refined to allow representation of stereochemical differences and charges as well as special kinds of bonding such as those seen in organo-metallic compounds. The principal advantage of a computer representation is the possibility for increased storage and fast, flexible search.

Search

Substructure

Chemists can search databases using parts of structures, parts of their IUPAC names as well as based on constraints on properties. Chemical databases are particularly different from other general purpose databases in their support for sub-structure search. This kind of search is achieved by looking for subgraph isomorphism (sometimes also called a monomorphism) and is a widely studied application of Graph theory. The algorithms for searching are computationally intensive, often of O (n^3) or O (n^4) time complexity (where n is the number of atoms involved). The intensive component of search is called atom-by-atom-searching (ABAS), in which a mapping of the search substructure atoms and bonds with the target molecule is sought. ABAS searching usually makes use of the Ullman algorithm or variations of it (*i.e.* SMSD). Speedups are achieved by time amortization, that is, some of the time on search tasks are saved by using precomputed information. This pre-computation typically involves creation of bitstrings representing presence or absence of molecular fragments. By looking at the fragments present in a search structure it is possible to eliminate the need for ABAS comparison with target molecules that do not possess the fragments that are present in the search structure. This elimination is called screening (differen than screening procedures used in drug-discovery). The bit-strings used for these applications are also called structural-keys. The performance of such keys depends on the choice of the fragments used for constructing the keys and the probability of their presence in the database molecules. Another kind of key makes use of hash-codes based on fragments derived computationally. These are called 'fingerprints' although the term is sometimes used synonymously with structural-keys. The amount of memory needed to store these structural-keys and fingerprints can be reduced by 'folding', which is achieved by combining parts of the key using bitwise-operations and thereby reducing the overall length.

Conformation

Search by matching 3D conformation of molecules or by specifying spatial constraints is another feature that is particularly of use in drug design. Searches of this kind can be computationally very expensive. Many approximate methods have been proposed, for instance BCUTS, special function representations, moments of inertia, ray-tracing histograms, maximum distance histograms, shape multipoles to name a few.

Descriptors

All properties of molecules beyond their structure can be split up into either physico-chemical or pharmacological attributes also called descriptors. On top of that, there exist various artificial and more or less standardized naming systems for molecules that supply more or less ambiguous names and synonyms. The IUPAC name is usually a good choice for representing a molecule's structure in a both human-readable and unique string although it becomes unwieldy for larger molecules. Trivial names on the other hand abound with homonyms and synonyms and are therefore a bad choice as a defining database key. While physico-chemical descriptors like molecular weight, (partial) charge, solubility, etc. can mostly be computed directly based on the molecule's structure, pharmacological descriptors can be derived only indirectly using involved multivariate statistics or experimental (screening, bioassay) results. All of those descriptors can for reasons of computational effort be stored along with the molecule's representation and usually are.

Similarity

There is no single definition of molecular similarity, however the concept may be defined according to the application and is often described as an inverse of a measure of distance in descriptor space. Two molecules might be considered more similar for instance if their difference in molecular weights is lower than when compared with others. A variety of other measures could be combined to produce a multi-variate distance measure. Distance measures are often classified into Euclidean measures and non-Euclidean measures depending on whether the triangle inequality holds. Maximum Common Subgraph (MCS) based substructure search (similarity or distance measure) is also very common. MCS is also used for screening drug like compounds by hitting molecules, which share common subgraph (substructure).

Chemicals in the databases may be clustered into groups of 'similar' molecules based on similarities. Both hierarchical and non-hierarchical clustering approaches can be applied to chemical entities with multiple attributes. These attributes or molecular properties may either be determined empirically or computationally derived descriptors. One of the most popular clustering approaches is the Jarvis-Patrick algorithm.

In pharmacologically oriented chemical repositories, similarity is usually defined in terms of the biological effects of compounds (ADME/tox) that can in turn be semiautomatically inferred from similar combinations of physico-chemical descriptors using QSAR methods.

Registration Systems

Databases systems for maintaining unique records on chemical compounds are termed as Registration systems. These are often used for chemical indexing, patent systems and industrial databases.

Registration systems usually enforce uniqueness of the chemical represented in the database through the use of unique representations. By applying rules of precedence for the generation of stringified notations, one can obtain unique/'canonical' string representations such as 'canonical SMILES'. Some registration systems such as the CAS system make use of algorithms to generate unique hash codes to achieve the same objective.

A key difference between a registration system and a simple chemical database is the ability to accurately represent that which is known, unknown, and partially known. For example, a chemical database might store a molecule with stereochemistry unspecified, whereas a chemical registry system requires the registrar to specify whether the stereo configuration is unknown, a specific (known) mixture, or racemic. Each of these would be considered a different record in a chemical registry system.

Registration systems also preprocess molecules to avoid considering trivial differences such as differences in halogen ions in chemicals.

An example is the Chemical Abstracts Service (CAS) registration system.

Tools

The computational representations are usually made transparent to chemists by graphical display of the data. Data entry is also simplified through the use of chemical structure editors. These editors internally convert the graphical data into computational representations.

There are also numerous algorithms for the interconversion of various formats of representation. An open-source utility for conversion is OpenBabel. These search and conversion algorithms are implemented either within the database system itself or as is now the trend is implemented as external components that fit into standard relational database systems. Both Oracle and PostgreSQL based systems make use of cartridge technology that allows user defined datatypes. These allow the user to make SQL queries with chemical search conditions (For example, a query to search for records having a phenyl ring in their structure represented as a SMILES string in a SMILESCOL column could be:

```
SELECT * FROM CHEMTABLE WHERE SMILESCOL.CONTAINS('c1ccccc1')
```

Algorithms for the conversion of IUPAC names to structure representations and vice versa are also used for extracting structural information from text. However, there are difficulties due to the existence of multiple dialects of IUPAC. Work is on to establish a unique IUPAC standard.

Chemical File Format

This topic discusses some common molecular file formats, including usage and converting between them.

Distinguishing Formats

Chemical information is usually provided as files or streams and many formats have been created, with varying degrees of documentation. The format is indicated in three ways:

- *file extension* (usually 3 letters). This is widely used, but fragile as common suffixes such as ".mol" and ".dat" are used by many systems, including non-chemical ones.

- *self-describing files* where the format information is included in the file. Examples are CIF and CML.

- *chemical/MIME type* added by a chemically-aware server.

Chemical Markup Language

Chemical Markup Language (CML) is an open standard for representing molecular and other chemical data. The open source project includes XML Schema, source code for parsing and working with CML data, and an active community. CML data files are accepted by many tools, including JChemPaint, Jmol, XDrawChem and MarvinView.

Protein Data Bank Format

The Protein Data Bank Format is commonly used for proteins but it can be used for other types of molecules as well. It was originally designed as, and continues to be, a fixed-column-width format and thus officially has a built-in maximum number of atoms, of residues, and of chains; this resulted in splitting very large structures such as ribosomes into multiple files. However, many tools can read files that exceed those limits. For example, the E. coli 70S ribosome was represented as 4 PDB files in 2009: 3I1M, 3I1N, 3I1O and 3I1P. In 2014 they were consolidated into a single file, 4V6C.

Some PDB files contain an optional section describing atom connectivity as well as position. Because these files are sometimes used to describe macromolecular assemblies or molecules represented in explicit solvent, they can grow very large and are often compressed. Some tools, such as Jmol and KiNG, can read PDB files in gzipped format. The wwPDB maintains the specifications of the PDB file format and its XML alternative, PDBML. There was a fairly major change in PDB format specification (to version 3.0) in August 2007, and a remediation of many file problems in the existing database. The typical file extension for a PDB file is *.pdb*, although some older files use *.ent* or *.brk*. Some molecular modeling tools write nonstandard PDB-style files that adapt the basic format to their own needs.

GROMACS Format

The GROMACS file format family was created for use with the molecular simulation software package GROMACS. It closely resembles the PDB format but was designed for storing output from molecular dynamics simulations, so it allows for additional numerical precision and optionally retains information about particle velocity as well as position at a given point in the simulation trajectory. It does not allow for the storage of connectivity information, which in GROMACS is obtained from separate molecule and system topology files. The typical file extension for a GROMACS file is *.gro*.

CHARMM Format

The CHARMM molecular dynamics package can read and write a number of standard chemical and biochemical file formats; however, the CARD (coordinate) and PSF (protein structure file) are largely unique to CHARMM. The CARD format is fixed-column-width, resembles the PDB format, and is used exclusively for storing atomic coordinates. The PSF file contains atomic connectivity information (which describes atomic bonds) and is required before beginning a simulation. The typical file extensions used are *.crd* and *.psf* respectively.

GSD Format

The General Simulation Data (GSD) file format created for efficient reading / writing of generic particle simulations, primarily - but not restricted to - those from HOOMD-blue. The package also contains a python module that reads and writes hoomd schema gsd files with an easy to use syntax.

Ghemical File Format

The Ghemical software can use OpenBabel to import and export a number of file formats. However, by default, it uses the GPR format. This file is composed of several parts, separated by a tag (!Header, !Info, !Atoms, !Bonds, !Coord, !PartialCharges and !End).

The proposed MIME type for this format is *application/x-ghemical*.

SYBYL Line Notation

SYBYL Line Notation (SLN) is a chemical line notation. Based on SMILES, it incorporates a complete syntax for specifying relative stereochemistry. SLN has a rich query syntax that allows for the specification of Markush queries. The syntax also supports the specification of combinatorial libraries of CD.

Example SLNs

Description	SLN String
Benzene	CH:CH:CH:CH:CH:CH:@1
Alanine	NH2C[s=n]H(CH3)C(=O)OH
Query showing R sidechain	R1[hac>1]C:C:C:C:C:C:@1
Query for amide/sulfamide	NHC=M1{M1:O,S}

SMILES

The Simplified Molecular Input Line Entry Specification (SMILES) is a line notation for molecules. SMILES strings include connectivity but do not include 2D or 3D coordinates.

Hydrogen atoms are not represented. Other atoms are represented by their element symbols B, C, N, O, F, P, S, Cl, Br, and I. The symbol "=" represents double bonds and "#" represents triple bonds. Branching is indicated by (). Rings are indicated by pairs of digits.

Some examples are

Name	Formula	SMILES String
Methane	CH_4	C
Ethanol	C_2H_6O	CCO
Benzene	C_6H_6	C1=CC=CC=C1 or c1ccccc1
Ethylene	C_2H_4	C=C

XYZ

The XYZ file format is a simple format that usually gives the number of atoms in the first line, a comment on the second, followed by a number of lines with atomic symbols (or atomic numbers) and cartesian coordinates.

MDL Number

The MDL number contains a unique identification number for each reaction and variation. The format is RXXXnnnnnnnn. R indicates a reaction, XXX indicates which database contains the reaction record. The numeric portion, nnnnnnnn, is an 8-digit number.

Other Common Formats

One of the most widely used industry standards are chemical table file formats, like the *Structure Data Format* (SDF) files. They are text files that adhere to a strict format for representing multiple chemical structure records and associated data fields. The format was originally developed and published by Molecular Design Limited (MDL). MOL is another file format from MDL. It is documented in Chapter 4 of *CTfile Formats*.

PubChem also has XML and ASN1 file formats, which are export options from the PubChem online database. They are both text based (ASN1 is most often a binary format).

There are a large number of other formats listed in the table below.

Converting Between Formats

OpenBabel and JOELib are freely available open source tools specifically designed for converting between file formats. Their chemical expert systems support a large atom type conversion tables.

```
babel -i input_format input_file -o output_format output_file
```

For example, to convert the file epinephrine.sdf in SDF to CML use the command

```
babel -i sdf epinephrine.sdf -o cml epinephrine.cml
```

The resulting file is epinephrine.cml.

A number of tools intended for viewing and editing molecular structures are able to read in files in a number of formats and write them out in other formats. The tools JChemPaint (based on the Chemistry Development Kit), XDrawChem (based on OpenBabel), Chime, Jmol, Mol2mol and Discovery Studio fit into this category.

The Chemical MIME Project

"Chemical MIME" is a de facto approach for adding MIME types to chemical streams.

This project started in January 1994, and was first announced during the Chemistry workshop at the First WWW International Conference, held at CERN in May 1994. The first version of an Internet draft was published during May–October 1994, and the second revised version during April–September 1995. A paper presented to the CPEP (Committee on Printed and Electronic Publications) at the IUPAC meeting in August 1996 is available for discussion.

In 1998 the work was formally published in the JCIM.

File Extension	MIME Type	Proper Name	Description
alc	chemical/x-alchemy	Alchemy Format	

csf	chemical/x-cache-csf	CAChe MolStruct CSF	
cbin, cascii, ctab	chemical/x-cactvs-binary	CACTVS format	
cdx	chemical/x-cdx	ChemDraw eXchange file	
cer	chemical/x-cerius	MSI Cerius II format	
c3d	chemical/x-chem3d	Chem3D Format	
chm	chemical/x-chemdraw	ChemDraw file	
cif	chemical/x-cif	Crystallographic Information File, Crystallographic Information Framework	Promulgated by the International Union of Crystallography
cmdf	chemical/x-cmdf	CrystalMaker Data format	
cml	chemical/x-cml	Chemical Markup Language	XML based Chemical Markup Language.
cpa	chemical/x-compass	Compass program of the Takahashi	
bsd	chemical/x-crossfire	Crossfire file	
csm, csml	chemical/x-csml	Chemical Style Markup Language	
ctx	chemical/x-ctx	Gasteiger group CTX file format	
cxf, cef	chemical/x-cxf	Chemical eXchange Format	
emb, embl	chemical/x-embl-dl-nucleotide	EMBL Nucleotide Format	
spc	chemical/x-galactic-spc	SPC format for spectral and chromatographic data	
inp, gam, gamin	chemical/x-gamess-input	GAMESS Input format	
fch, fchk	chemical/x-gaussian-checkpoint	Gaussian Checkpoint Format	
cub	chemical/x-gaussian-cube	Gaussian Cube (Wavefunction) Format	
gau, gjc, gjf, com	chemical/x-gaussian-input	Gaussian Input Format	
gcg	chemical/x-gcg8-sequence	Protein Sequence Format	
gen	chemical/x-genbank	ToGenBank Format	
istr,ist	chemical/x-isostar	IsoStar Library of Intermolecular Interactions	
jdx, dx	chemical/x-jcamp-dx	JCAMP Spectroscopic Data Exchange Format	
kin	chemical/x-kinemage	Kinetic (Protein Structure) Images; Kinemage	
mcm	chemical/x-macmolecule	MacMolecule File Format	
mmd, mmod	chemical/x-macromodel-input	MacroModel Molecular Mechanics	

mol	chemical/x-mdl-molfile	MDL Molfile	
smiles, smi	chemical/x-day-light-smiles	Simplified molecular input line entry specification	A line notation for molecules.
sdf	chemical/x-mdl-sdfile	Structure-Data File	
el	chemical/x-sketchel	SketchEl Molecule	
ds	chemical/x-datasheet	SketchEl XML DataSheet	
inchi	chemical/x-inchi	The IUPAC International Chemical Identifier	
jsd, jsdraw	chemical/x-jsdraw	JSDraw native file format	
helm, ihelm	chemical/x-helm	Pistoia Alliance HELM string	A line notation for biological molecules
xhelm	chemical/x-xhelm	Pistoia Alliance XHELM XML file	XML based HELM including monomer definitions

Support

For Linux/Unix, configuration files are available as a *"chemical-mime-data"* package in .deb, RPM and tar.gz formats to register chemical MIME types on a web server. Programs can then register as viewer, editor or processor for these formats so that full support for chemical MIME types is available.

Sources of Chemical Data

Here is a short list of sources of freely available molecular data. There are many more resources than listed here out there on the Internet.

1. The US National Institute of Health PubChem database is a huge source of chemical data. All of the data is in two-dimensions. Data includes SDF, SMILES, PubChem XML, and PubChem ASN1 formats.

2. The worldwide Protein Data Bank (wwPDB) is an excellent source of protein and nucleic acid molecular coordinate data. The data is three-dimensional and provided in Protein Data Bank (PDB) format.

3. eMolecules is a commercial database for molecular data. The data includes a two-dimensional structure diagram and a smiles string for each compound. eMolecules supports fast substructure searching based on parts of the molecular structure.

4. ChemExper is a commercial data base for molecular data. The search results include a two-dimensional structure diagram and a mole file for many compounds.

5. New York University Library of 3-D Molecular Structures.

6. The US Environmental Protection Agency's The Distributed Structure-Searchable Toxicity (DSSTox) Database Network is a project of EPA's Computational Toxicology Program. The database provides SDF molecular files with a focus on carcinogenic and otherwise toxic substances.

Quantitative Structure–activity Relationship

Quantitative structure–activity relationship models (QSAR models) are regression or classification models used in the chemical and biological sciences and engineering. Like other regression models, QSAR regression models relate a set of "predictor" variables (X) to the potency of the response variable (Y), while classification QSAR models relate the predictor variables to a categorical value of the response variable.

In QSAR modeling, the predictors consist of physico-chemical properties or theoretical molecular descriptors of chemicals; the QSAR response-variable could be a biological activity of the chemicals. QSAR models first summarize a supposed relationship between chemical structures and biological activity in a data-set of chemicals. Second, QSAR models predict the activities of new chemicals.

Related terms include *quantitative structure–property relationships (QSPR)* when a chemical property is modeled as the response variable. "Different properties or behaviors of chemical molecules have been investigated in the field of QSPR. Some examples are quantitative structure–reactivity relationships (QSRRs), quantitative structure–chromatography relationships (QSCRs) and, quantitative structure–toxicity relationships (QSTRs), quantitative structure–electrochemistry relationships (QSERs), and quantitative structure–biodegradability relationships (QSBRs)."

As an example, biological activity can be expressed quantitatively as the concentration of a substance required to give a certain biological response. Additionally, when physicochemical properties or structures are expressed by numbers, one can find a mathematical relationship, or quantitative structure-activity relationship, between the two. The mathematical expression, if carefully validated can then be used to predict the modeled response of other chemical structures.

A QSAR has the form of a mathematical model:

*   Activity = f(physiochemical properties and/or structural properties) + error

The error includes model error (bias) and observational variability, that is, the variability in observations even on a correct model.

Essential Steps in QSAR Studies

Principal steps of QSAR/QSPR including (i) Selection of Data set and extraction of structural/empirical descriptors (ii) variable selection, (iii) model construction and (iv) validation evaluation.

SAR and the SAR Paradox

The basic assumption for all molecule based hypotheses is that similar molecules have similar activities. This principle is also called Structure–Activity Relationship (SAR). The underlying problem is therefore how to define a *small* difference on a molecular level, since each kind of activity, e.g. reaction ability, biotransformation ability, solubility, target activity, and so on, might depend on another difference. Good examples were given in the bioisosterism reviews by Patanie/LaVoie and Brown.

In general, one is more interested in finding strong trends. Created hypotheses usually rely on a finite number of chemical data. Thus, the induction principle should be respected to avoid overfitted hypotheses and deriving overfitted and useless interpretations on structural/molecular data.

The *SAR paradox* refers to the fact that it is not the case that all similar molecules have similar activities.

Types

Fragment Based (Group Contribution)

Analogously, the "partition coefficient"—a measurement of differential solubility and itself a component of QSAR predictions—can be predicted either by atomic methods (known as "XLogP" or "ALogP") or by chemical fragment methods (known as "CLogP" and other variations). It has been shown that the logP of compound can be determined by the sum of its fragments; fragment-based methods are generally accepted as better predictors than atomic-based methods. Fragmentary values have been determined statistically, based on empirical data for known logP values. This method gives mixed results and is generally not trusted to have accuracy of more than ±0.1 units.

Group or Fragment based QSAR is also known as GQSAR. GQSAR allows flexibility to study various molecular fragments of interest in relation to the variation in biological response. The molecular fragments could be substituents at various substitution sites in congeneric set of molecules or could be on the basis of pre-defined chemical rules in case of non-congeneric sets. GQSAR also considers cross-terms fragment descriptors, which could be helpful in identification of key fragment interactions in determining variation of activity. Lead discovery using Fragnomics is an emerging paradigm. In this context FB-QSAR proves to be a promising strategy for fragment library design and in fragment-to-lead identification endeavours.

An advanced approach on fragment or group-based QSAR based on the concept of pharmacophore-similarity is developed. This method, pharmacophore-similarity-based QSAR (PS-QSAR) uses topological pharmacophoric descriptors to develop QSAR models. This activity prediction may assist the contribution of certain pharmacophore features encoded by respective fragments toward activity improvement and/or detrimental effects.

3D-QSAR

The acronym 3D-QSAR or 3-D QSAR refers to the application of force field calculations requiring three-dimensional structures of a given set of small molecules with known activities (training set). The training set need to be superimposed (aligned) by either experimental data (e.g. based on ligand-protein crystallography) or molecule superimposition software. It uses computed potentials, e.g. the Lennard-Jones potential, rather than experimental constants and is concerned with the overall molecule rather than a single substituent. The first 3-D QSAR was named Comparative Molecular Field Analysis (CoMFA) by Cramer et al. It examined the steric fields (shape of the molecule) and the electrostatic fields which were correlated by means of partial least squares regression (PLS).

The created data space is then usually reduced by a following feature extraction. The following learning method can be any of the already mentioned machine learning methods, e.g. support vector machines. An alternative approach uses multiple-instance learning by encoding molecules as sets of data instances, each of which represents a possible molecular conformation. A label or response is assigned to each set corresponding to the activity of the molecule, which is assumed to be determined by at least one instance in the set (i.e. some conformation of the molecule).

On June 18, 2011 the Comparative Molecular Field Analysis (CoMFA) patent has dropped any restriction on the use of GRID and partial least-squares (PLS) technologies and the Rome Center for Molecular Design (RCMD) team opened a 3-D QSAR web server. Recently (October 2016) the 3D QSAR web server has been updated and opened to the public four basic web applications: Py-MolEdit, Py-ConfSearch, Py-Align an Py-CoMFA. The suffix Py stands for python as both the web site and the application have been developed with the python language. The four applications allow to build a 3-D QSAR model from scratch by simply knowing the training set structures and bioactivities. The 3D-QSAR server includes all the features to analyze the molecular interactions fields (MIFs) and all the 3-D QSAR maps in a 3-D fashion and interactive way.

Chemical Descriptor Based

In this approach, descriptors quantifying various electronic, geometric, or steric properties of a molecule are computed and used to develop a QSAR. This approach is different from the fragment (or group contribution) approach in that the descriptors are computed for the system as whole rather than from the properties of individual fragments. This approach is different from the 3D-QSAR approach in that the descriptors are computed from scalar quantities (e.g., energies, geometric parameters) rather than from 3D fields.

An example of this approach is the QSARs developed for olefin polymerization by half sandwich compounds.

Modeling

In the literature it can be often found that chemists have a preference for partial least squares (PLS) methods, since it applies the feature extraction and induction in one step.

Data Mining Approach

Computer SAR models typically calculate a relatively large number of features. Because those lack structural interpretation ability, the preprocessing steps face a feature selection problem (i.e., which structural features should be interpreted to determine the structure-activity relationship). Feature selection can be accomplished by visual inspection (qualitative selection by a human); by data mining; or by molecule mining.

A typical data mining based prediction uses e.g. support vector machines, decision trees, neural networks for inducing a predictive learning model.

Molecule mining approaches, a special case of structured data mining approaches, apply a similarity matrix based prediction or an automatic fragmentation scheme into molecular substructures. Furthermore, there exist also approaches using maximum common subgraph searches or graph kernels.

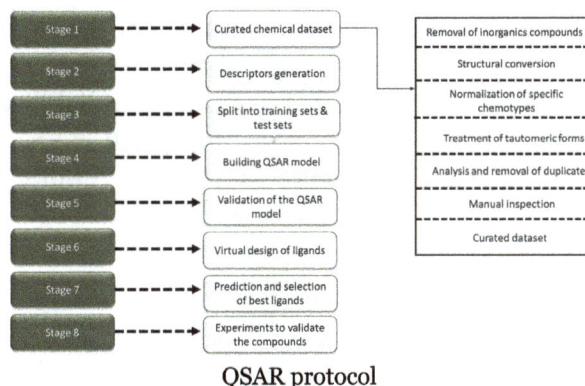

QSAR protocol

Matched Molecular Pair Analysis

Typically QSAR models derived from non linear machine learning is seen as a "black box", which fails to guide medicinal chemists. Recently there is a relatively new concept of matched molecular pair analysis or prediction driven MMPA which is coupled with QSAR model in order to identify activity cliffs.

Evaluation of the Quality of QSAR Models

QSAR modeling produces predictive models derived from application of statistical tools correlating biological activity (including desirable therapeutic effect and undesirable side effects) or physico-chemical properties in QSPR models of chemicals (drugs/

toxicants/environmental pollutants) with descriptors representative of molecular structure or properties. QSARs are being applied in many disciplines, for example: risk assessment, toxicity prediction, and regulatory decisions in addition to drug discovery and lead optimization. Obtaining a good quality QSAR model depends on many factors, such as the quality of input data, the choice of descriptors and statistical methods for modeling and for validation. Any QSAR modeling should ultimately lead to statistically robust and predictive models capable of making accurate and reliable predictions of the modeled response of new compounds.

For validation of QSAR models, usually various strategies are adopted:

1. internal validation or cross-validation (actually, while extracting data, cross validation is a measure of model robustness, the more a model is robust (higher q2) the less data extraction perturb the original model);

2. external validation by splitting the available data set into training set for model development and prediction set for model predictivity check;

3. blind external validation by application of model on new external data and

4. data randomization or Y-scrambling for verifying the absence of chance correlation between the response and the modeling descriptors.

The success of any QSAR model depends on accuracy of the input data, selection of appropriate descriptors and statistical tools, and most importantly validation of the developed model. Validation is the process by which the reliability and relevance of a procedure are established for a specific purpose; for QSAR models validation must be mainly for robustness, prediction performances and applicability domain (AD) of the models.

Some validation methodologies can be problematic. For example, *leave one-out* cross-validation generally leads to an overestimation of predictive capacity. Even with external validation, it is difficult to determine whether the selection of training and test sets was manipulated to maximize the predictive capacity of the model being published.

Different aspects of validation of QSAR models that need attention includes methods of selection of training set compounds, setting training set size and impact of variable selection for training set models for determining the quality of prediction. Development of novel validation parameters for judging quality of QSAR models is also important.

Application

Chemical

One of the first historical QSAR applications was to predict boiling points.

It is well known for instance that within a particular family of chemical compounds, especially of organic chemistry, that there are strong correlations between structure and observed properties. A simple example is the relationship between the number of carbons in alkanes and their boiling points. There is a clear trend in the increase of boiling point with an increase in the number carbons, and this serves as a means for predicting the boiling points of higher alkanes.

A still very interesting application is the Hammett equation, Taft equation and pKa prediction methods.

Biological

The biological activity of molecules is usually measured in assays to establish the level of inhibition of particular signal transduction or metabolic pathways. Drug discovery often involves the use of QSAR to identify chemical structures that could have good inhibitory effects on specific targets and have low toxicity (non-specific activity). Of special interest is the prediction of partition coefficient log P, which is an important measure used in identifying "druglikeness" according to Lipinski's Rule of Five.

While many quantitative structure activity relationship analyses involve the interactions of a family of molecules with an enzyme or receptor binding site, QSAR can also be used to study the interactions between the structural domains of proteins. Protein-protein interactions can be quantitatively analyzed for structural variations resulted from site-directed mutagenesis.

It is part of the machine learning method to reduce the risk for a SAR paradox, especially taking into account that only a finite amount of data is available. In general, all QSAR problems can be divided into coding and learning.

Applications

(Q)SAR models have been used for risk management. QSARS are suggested by regulatory authorities; in the European Union, QSARs are suggested by the REACH regulation, where "REACH" abbreviates "Registration, Evaluation, Authorisation and Restriction of Chemicals".

The chemical descriptor space whose convex hull is generated by a particular training set of chemicals is called the training set's applicability domain. Prediction of properties of novel chemicals that are located outside the applicability domain uses extrapolation, and so is less reliable (on average) than prediction within the applicability domain. The assessment of the reliability of QSAR predictions remains a research topic.

The QSAR equations can be used to predict biological activities of newer molecules before their synthesis.

Examples of machine learning tools for QSAR modeling include:

S.No.	Name	Algorithms	External link
1.	R	RF, SVM, Naïve Bayesian, and ANN	*"R: The R Project for Statistical Computing".*
2.	libSVM	SVM	*"LIBSVM -- A Library for Support Vector Machines".*
3.	Orange	RF, SVM, and Naïve Bayesian	*"Orange Data Mining".*
4.	RapidMiner	SVM, RF, Naïve Bayes, DT, ANN, and k-NN	*"RapidMiner	#1 Open Source Predictive Analytics Platform".*
5.	Weka	RF, SVM, and Naïve Bayes	*"Weka 3 - Data Mining with Open Source Machine Learning Software in Java".*
6.	Knime	DT, Naïve Bayes, and SVM	*"KNIME	Open for Innovation".*
7.	AZOrange	RT, SVM, ANN, and RF	*"AZCompTox/AZOrange: AstraZeneca add-ons to Orange.". GitHub.*
8.	Tanagra	SVM, RF, Naïve Bayes, and DT	*"TANAGRA - A free DATA MINING software for teaching and research".*
9.	Elki	k-NN	*"ELKI Data Mining Framework".*
10.	MALLET		*"MALLET homepage".*
11.	MOA		*"MOA Massive Online Analysis	Real Time Analytics for Data Streams".*

References

- Ullmann, Julian R. (1976), "An algorithm for subgraph isomorphism", Journal of the ACM, 23 (1): 31–42, doi:10.1145/321921.321925

- Vert JP, Schölkopf B, Tsuda K (2004). Kernel methods in computational biology. Cambridge, Mass: MIT Press. ISBN 0-262-19509-7

- Alexandre Varnek and Igor Baskin (2011). "Chemoinformatics as a Theoretical Chemistry Discipline". Molecular Informatics. 30 (1): 20–32. doi:10.1002/minf.201000100

- Rahman, S. A.; Bashton, M.; Holliday, G. L.; Schrader, R.; Thornton, J. M. (2000). "Small Molecule Subgraph Detector (SMSD) toolkit". Journal of Cheminformatics. 1: 12. doi:10.1186/1758-2946-1-12

- Gusfield D (1997). Algorithms on strings, trees, and sequences: computer science and computational biology. Cambridge, UK: Cambridge University Press. ISBN 0-521-58519-8

- Henrick, K.; et al. (2008). "Remediation of the protein data bank archive". Nucleic Acids Research. 36 (Database issue): D426–D433. PMC 2238854. PMID 18073189. doi:10.1093/nar/gkm937

- Dearden JC (2003). "In silico prediction of drug toxicity". Journal of Computer-Aided Molecular Design. 17 (2-4): 119–27. Bibcode:2003JCAMD..17..119D. PMID 13677480. doi:10.1023/A:1025361621494

- Timmerman H, Todeschini R, Consonni V, Mannhold R, Kubinyi H (2002). Handbook of Molecular Descriptors. Weinheim: Wiley-VCH. ISBN 3-527-29913-0

- Berman, H.M.; et al. (2003). "Announcing the worldwide Protein Data Bank". Nature Structural Biology. 10 (12): 980. PMID 14634627. doi:10.1038/nsb1203-980

- Prasanth Kumar S, Jasrai YT, Pandya HA, Rawal RM (November 2013). "Pharmacophore-similarity-based QSAR (PS-QSAR) for group-specific biological activity predictions". Journal of Biomolecular Structure & Dynamics. 33 (1): 56–69. PMID 24266725. doi:10.1080/07391102.2013.849618

- Rouvray DH, Bonchev D (1991). Chemical graph theory: introduction and fundamentals. Tunbridge Wells, Kent, England: Abacus Press. ISBN 0-85626-454-7

- Pratim Roy P, Paul S, Mitra I, Roy K (2009). "On two novel parameters for validation of predictive QSAR models". Molecules. 14 (5): 1660–701. PMID 19471190. doi:10.3390/molecules14051660.

- Rahman, S. Asad; Bashton, M.; Holliday, G. L.; Schrader, R.; Thornton, J. M. (2009). "Small Molecule Subgraph Detector (SMSD) Toolkit". Journal of Cheminformatics. 1: 12. doi:10.1186/1758-2946-1-12

- Dietterich TG, Lathrop RH, Lozano-Pérez T (1997). "Solving the multiple instance problem with axis-parallel rectangles". Artificial Intelligence. 89 (1–2): 31–71. doi:10.1016/S0004-3702(96)00034-3

Methods Used in Computational Chemistry

The methods used in computational chemistry are ab initio quantum chemistry methods, quantum chemistry composite methods, Møller–Plesset perturbation theory etc. Ab initio quantum chemistry method is entirely based on quantum chemistry. The aspects elucidated in this section are of vital importance, and provide a better understanding of ab initio methods.

Ab initio Quantum Chemistry Methods

Ab initio quantum chemistry methods are computational chemistry methods based on quantum chemistry. The term *ab initio* was first used in quantum chemistry by Robert Parr and coworkers, including David Craig in a semiempirical study on the excited states of benzene. The background is described by Parr. In its modern meaning ('from first principles of quantum mechanics') the term was used by Chen (when quoting an unpublished 1955 MIT report by Allen and Nesbet), by Roothaan and, in the title of an article, by Allen and Karo, who also clearly define it.

Almost always the basis set (which is usually built from the LCAO ansatz) used to solve the Schrödinger equation is not complete, and does not span the Hilbert space associated with ionization and scattering processes. In the Hartree–Fock method and the configuration interaction method, this approximation allows one to treat the Schrödinger equation as a "simple" eigenvalue equation of the electronic molecular Hamiltonian, with a discrete set of solutions.

Accuracy and Scaling

Ab initio electronic structure methods have the advantage that they can be made to converge to the exact solution, when all approximations are sufficiently small in magnitude and when the finite set of basis functions tends toward the limit of a complete set. In this case, configuration interaction, where all possible configurations are included (called "Full CI"), tends to the exact non-relativistic solution of the electronic Schrödinger equation (in the Born–Oppenheimer approximation). The convergence, however, is usually not monotonic, and sometimes the smallest calculation gives the best result for some properties.

One needs to consider the computational cost of *ab initio* methods when determining whether they are appropriate for the problem at hand. When compared to much less accurate approaches, such as molecular mechanics, *ab initio* methods often take larger amounts of computer time, memory, and disk space, though, with modern advances in computer science and technology such considerations are becoming less of an issue. The HF method scales nominally as N^4 (N being a relative measure of the system size, not the number of basis functions) – e.g., if you double the number of electrons and the number of basis functions (double the system size), the calculation will take 16 (2^4) times as long per iteration. However, in practice it can scale closer to N^3 as the program can identify zero and extremely small integrals and neglect them. Correlated calculations scale less favorably, though their accuracy is usually greater, which is the trade off one needs to consider: Second–order many–body perturbation theory (MBPT(2)), or when the HF reference is used, Møller–Plesset perturbation theory (MP2) scales as N^4 or N^5, depending on how it is implemented, MP3 scales as N^6 and coupled cluster with singles and doubles (CCSD) scales iteratively as N^6, MP4 scales as N^7 and CCSD(T) and CR-CC(2,3) scale iteratively as N^6, with one noniterative step which scales as N^7. Density functional theory (DFT) methods using functionals which include Hartree–Fock exchange scale in a similar manner to Hartree–Fock but with a larger proportionality term and are thus more expensive than an equivalent Hartree–Fock calculation. DFT methods that do not include Hartree–Fock exchange can scale better than Hartree–Fock.

Linear Scaling Approaches

The problem of computational expense can be alleviated through simplification schemes. In the *density fitting* scheme, the four-index integrals used to describe the interaction between electron pairs are reduced to simpler two- or three-index integrals, by treating the charge densities they contain in a simplified way. This reduces the scaling with respect to basis set size. Methods employing this scheme are denoted by the prefix "df-", for example the density fitting MP2 is df-MP2 (many authors use lower-case to prevent confusion with DFT). In the *local approximation*, the molecular orbitals are first localized by a unitary rotation in the orbital space (which leaves the reference wave function invariant, i.e., is not an approximation) and subsequently interactions of distant pairs of localized orbitals are neglected in the correlation calculation. This sharply reduces the scaling with molecular size, a major problem in the treatment of biologically-sized molecules. Methods employing this scheme are denoted by the prefix "L", e.g. LMP2. Both schemes can be employed together, as in the df-LMP2 and df-LCCSD(T0) methods. In fact, df-LMP2 calculations are faster than df-Hartree–Fock calculations and thus are feasible in nearly all situations in which also DFT is.

Classes of Methods

The most popular classes of *ab initio* electronic structure methods:

Hartree–Fock Methods

- Hartree–Fock (HF)

- Restricted open-shell Hartree–Fock (ROHF)

- Unrestricted Hartree–Fock (UHF)

Post-Hartree–Fock Methods

- Møller–Plesset perturbation theory (MPn)

- Configuration interaction (CI)

- Coupled cluster (CC)

- Quadratic configuration interaction (QCI)

- Quantum chemistry composite methods

Multi-reference Methods

- Multi-configurational self-consistent field (MCSCF including CASSCF and RASSCF)

- Multi-reference configuration interaction (MRCI)

- n-electron valence state perturbation theory (NEVPT)

- Complete active space perturbation theory (CASPTn)

- State universal multi-reference coupled-cluster theory (SUMR-CC)

Methods in Detail

Hartree–Fock and Post-Hartree–Fock Methods

The simplest type of *ab initio* electronic structure calculation is the Hartree–Fock (HF) scheme, in which the instantaneous Coulombic electron-electron repulsion is not specifically taken into account. Only its average effect (mean field) is included in the calculation. This is a variational procedure; therefore, the obtained approximate energies, expressed in terms of the system's wave function, are always equal to or greater than the exact energy, and tend to a limiting value called the Hartree–Fock limit as the size of the basis is increased. Many types of calculations begin with a Hartree–Fock calculation and subsequently correct for electron-electron repulsion, referred to also as electronic correlation. Møller–Plesset perturbation theory (MPn) and coupled cluster theory (CC) are examples of these post-Hartree–Fock methods. In some cases, particularly for bond breaking processes, the Hartree–Fock method is inadequate and this single-determinant reference function is not a good basis for post-Hartree–Fock methods. It is then necessary to start with a wave function that includes more than one determinant such as multi-configurational self-consistent field (MCSCF) and methods have been developed that use these multi-determinant references for improvements. However, if one uses coupled cluster methods such as CCSDT, CCSDt, CR-CC(2,3), or CC(t;3) then single-bond breaking using the single determinant HF reference is feasible. For an accurate description of double bond breaking, methods such as CCSDTQ, CCSDTq, CCSDtq, CR-CC(2,4), or CC(tq;3,4) also make use of the single determinant HF reference, and do not require one to use multi-reference methods.

Example

Is the bonding situation in disilyne Si_2H_2 the same as in acetylene (C_2H_2) ?

A series of *ab initio* studies of Si_2H_2 is an example of how *ab initio* computational chemistry can predict new structures that are subsequently confirmed by experiment. They go back over 20 years, and most of the main conclusions were reached by 1995. The methods used were mostly post-Hartree–Fock, particularly configuration interaction (CI) and coupled cluster (CC). Initially the question was whether disilyne, Si_2H_2 had the same structure as ethyne (acetylene), C_2H_2. In early studies, by Binkley and Lischka and Kohler, it became clear that linear Si_2H_2 was a transition structure between two equivalent trans-bent structures and that the ground state was predicted to be a four-membered ring bent in a 'butterfly' structure with hydrogen atoms bridged between the two silicon atoms. Interest then moved to look at whether structures equivalent to vinylidene ($Si=SiH_2$) existed. This structure is predicted to be a local minimum, i. e. an isomer of Si_2H_2, lying higher in energy than the ground state but below the energy of the trans-bent isomer. Then a new isomer with an unusual structure was predicted by Brenda Colegrove in Henry F. Schaefer, III's group. It requires post-Hartree–Fock methods to obtain a local minimum for this structure. It does not exist on the Hartree–Fock energy hypersurface. The new isomer is a planar structure with one bridging hydrogen atom and one terminal hydrogen atom, cis to the bridging atom. Its energy is above the ground state but below that of the other isomers. Similar results were later obtained for Ge_2H_2. Al_2H_2 and Ga_2H_2 have exactly the same isomers, in spite of having two electrons less than the Group 14 molecules. The only difference is that the four-membered ring ground state is planar and not bent. The cis-mono-bridged and vinylidene-like isomers are present. Experimental work on these molecules is not easy, but matrix isolation spectroscopy of the products of the reaction of hydrogen atoms and silicon and aluminium surfaces has found the ground state ring structures and the cis-mono-bridged structures for Si_2H_2 and Al_2H_2. Theoretical predictions of the vibrational frequencies were crucial in understanding the experimental observations of the spectra of a mixture of compounds. This may appear to be an obscure area of chemistry, but the differences between carbon and silicon chemistry is always a lively question, as are the differences between group 13 and group 14 (mainly the B and C differences). The silicon and germanium compounds were the subject of a Journal of Chemical Education article.

Valence Bond Methods

Valence bond (VB) methods are generally *ab initio* although some semi-empirical versions have been proposed. Current VB approaches are:-

- Generalized valence bond (GVB)

- Modern valence bond theory (MVBT)

Quantum Monte Carlo Methods

A method that avoids making the variational overestimation of HF in the first place is Quantum Monte Carlo (QMC), in its variational, diffusion, and Green's function forms. These methods work with an explicitly correlated wave function and evaluate integrals numerically using a Monte Carlo integration. Such calculations can be very time-consuming. The accuracy of QMC depends strongly on the initial guess of many-body wave-functions and the form of the many-body wave-function. One simple choice is Slater-Jastrow wave-function in which the local correlations are treated with the Jastrow factor.

Coupled Cluster

Coupled cluster (CC) is a numerical technique used for describing many-body systems. Its most common use is as one of several post-Hartree–Fock ab initio quantum chemistry methods in the field of computational chemistry, but it is also used in nuclear physics. Coupled cluster essentially takes the basic Hartree–Fock molecular orbital method and constructs multi-electron wavefunctions using the exponential cluster operator to account for electron correlation. Some of the most accurate calculations for small to medium-sized molecules use this method.

The method was initially developed by Fritz Coester and Hermann Kümmel in the 1950s for studying nuclear physics phenomena, but became more frequently used when in 1966 Jiří Čížek (and later together with Josef Paldus) reformulated the method for electron correlation in atoms and molecules. It is now one of the most prevalent methods in quantum chemistry that includes electronic correlation. CC theory is simply the perturbative variant of the Many Electron Theory (MET) of Oktay Sinanoğlu, which is the exact (and variational) solution of the many electron problem, so it was also called "Coupled Pair MET (CPMET)". J. Čížek used the correlation function of MET and used Goldstone type perturbation theory to get the energy expression while original MET was completely variational. Čížek first developed the Linear-CPMET and then generalized it to full CPMET in the same paper in 1966. He then also performed an application of it on the benzene molecule with O. Sinanoğlu in the same year. Because MET is somewhat difficult to perform computationally, CC is simpler and thus, in today's computational chemistry, CC is the best variant of MET and gives highly accurate results in comparison to experiments.

Wavefunction Ansatz

Coupled-cluster theory provides the exact solution to the time-independent Schrödinger equation

$$H \,|\, \Psi \rangle = E \,|\, \Psi \rangle$$

where H is the Hamiltonian of the system, $|\Psi\rangle$ the exact wavefunction, and E the exact energy of the ground state. Coupled-cluster theory can also be used to obtain solutions for excited states using, for example, linear-response, equation-of-motion, state-universal multi-reference coupled cluster, or valence-universal multi-reference coupled cluster approaches.

The wavefunction of the coupled-cluster theory is written as an exponential ansatz:

$$|\Psi\rangle = e^{T} \,|\, \Phi_0 \rangle,$$

where $|\Phi_0\rangle$, the reference wave function, which is typically a Slater determinant constructed from Hartree–Fock molecular orbitals, though other wave functions such as configuration interaction, multi-configurational self-consistent field, or Brueckner orbitals can also be used. T is the cluster operator which, when acting on $|\Phi_0\rangle$, produces a linear combination of excited determinants from the reference wave function.

The choice of the exponential ansatz is opportune because (unlike other ansatzes, for example, configuration interaction) it guarantees the size extensivity of the solution. Size consistency in CC theory, also unlike other theories, does not depend on the size consistency of the reference wave function. This is easily seen, for example, in the single bond breaking of F_2 when using a restricted Hartree-Fock (RHF) reference, which is not size consistent, at the CCSDT level of theory which provides an almost exact, full CI-quality, potential energy surface and does not dissociate the molecule into F^- and F^+ ions, like the RHF wave function, but rather into two neutral F atoms. If one were to use, for example, the CCSD, CCSD[T], or CCSD(T) levels of theory, they would not provide reasonable results for the bond breaking of F_2, with the latter two approaches providing unphysical potential energy surfaces, though this is for reasons other than just size consistency.

A criticism of the method is that the conventional implementation employing the similarity-transformed Hamiltonian is not variational, though there are bi-variational and quasi-variational approaches that have been developed since the first implementations of the theory. While the above ansatz for the wave function itself has no natural truncation, however, for other properties, such as energy, there is a natural truncation when examining expectation values, which has its basis in the linked- and connected-cluster theorems, and thus does not suffer from issues such as lack of size extensivity, like the variational configuration interaction approach.

Cluster Operator

The cluster operator is written in the form,

$$T = T_1 + T_2 + T_3 + \cdots,$$

where T_1 is the operator of all single excitations, T_2 is the operator of all double excitations and so forth. In the formalism of second quantization these excitation operators are expressed as

$$T_1 = \sum_i \sum_a t_a^i \hat{a}^a \hat{a}_i,$$

$$T_2 = \frac{1}{4} \sum_{i,j} \sum_{a,b} t_{ab}^{ij} \hat{a}^a \hat{a}^b \hat{a}_j \hat{a}_i,$$

and for the general n-fold cluster operator

$$T_n = \frac{1}{(n!)^2} \sum_{i_1,i_2,\dots,i_n} \sum_{a_1,a_2,\dots,a_n} t_{a_1,a_2,\dots,a_n}^{i_1,i_2,\dots,i_n} \hat{a}^{a_1} \hat{a}^{a_2} \dots \hat{a}^{a_n} \hat{a}_{i_n} \dots \hat{a}_{i_2} \hat{a}_{i_1}.$$

In the above formulae $(\hat{a}_a^\dagger =)\hat{a}^a$ and $\hat{a}_i$ denote the creation and annihilation operators, respectively, and i, j stand for occupied (hole) and a, b for unoccupied (particle) orbitals (states). The creation and annihilation operators in the coupled cluster terms above are written in canonical form, where each term is in the normal order form, with respect to the Fermi vacuum, $|\Phi_0\rangle$. Being the one-particle cluster operator and the two-particle cluster operator, T_1 and T_2 convert the reference function $|\Phi_0\rangle$ into a linear combination of the singly and doubly excited Slater determinants, respectively, if applied without the exponential (such as in CI where a linear excitation operator is applied to the wave function). Applying the exponential cluster operator to the wave function, one can then generate more than doubly excited determinants due to the various powers of T_1 and T_2 that appear in the resulting expressions. Solving for the unknown coefficients t_a^i and t_{ab}^{ij} is necessary for finding the approximate solution $|\Psi\rangle$.

The exponential operator e^T may be expanded as a Taylor series and if we consider only the T_1 and T_2 cluster operators of T, we can write:

$$e^T = 1 + T + \frac{1}{2!}T^2 + \cdots = 1 + T_1 + T_2 + \frac{1}{2}T_1^2 + T_1 T_2 + \frac{1}{2}T_2^2 + \cdots$$

Though this series is finite in practice because the number of occupied molecular orbitals is finite, as is the number of excitations, it is still very large, to the extent that even modern day massively parallel computers are inadequate, except for problems of a dozen or so electrons and very small basis sets, when considering all contributions to the cluster operator and not just T_1 and T_2. Often, as was done above, the cluster operator includes only singles and doubles as this offers a computationally affordable method that performs better than MP2 and CISD, but is not very accurate usually. For accurate results some form of triples (approximate or full) are needed, even near the equilibrium geometry (in the Franck-Condon region), and especially when breaking single-bonds or describing diradical species (these latter examples are often what is referred to as multi-reference problems, since more than one determinant has a significant contribution to the resulting wave function). For double bond breaking, and more complicated problems in chemistry, quadruple excitations often become important as well, though usually they are small for most problems, and as such, the contribution of T_5, T_6 etc. to the operator T is typically small. Furthermore, if the highest excitation level in the T operator is n,

$$T = T_1 + ... + T_n$$

then Slater determinants for an N-electron system excited more than n $(< N)$ times may still contribute to the coupled cluster wave function $|\Psi\rangle$ because of the non-linear nature of the exponential ansatz, and therefore, coupled cluster terminated at T_n usually recovers more correlation energy than CI with maximum n excitations.

Coupled-cluster Equations

The Schrödinger equation can be written, using the coupled-cluster wave function, as

$$H|\Psi_0\rangle = He^T|\Phi_0\rangle = Ee^T|\Phi_0\rangle$$

where there are a total of q coefficients (t-amplitudes) to solve for. To obtain the q equations, first, we multiply the above Schrödinger equation on the left by e^{-T} and then project onto the entire set of up to m-tuply excited determinants, where m is the highest order excitation included in T, that can be constructed from the reference wave function $|\Phi_0\rangle$, denoted by $|\Phi^*\rangle$, and individually, $|\Phi_i^a\rangle$ are singly excited determinants where the electron in orbital i has been excited to orbital a; $|\Phi_{ij}^{ab}\rangle$ are doubly excited determinants where the electron in orbital i has been excited to orbital a and the electron in orbital j has been excited to orbital b, etc. In this way we generate a set of coupled energy-independent non-linear algebraic equations needed to determine the t-amplitudes.

$$\langle\Phi_0|e^{-T}He^T|\Phi_0\rangle = E\langle\Phi_0|\Phi_0\rangle = E$$

$$\langle\Phi^*|e^{-T}He^T|\Phi_0\rangle = E\langle\Phi^*|\Phi_0\rangle = 0,$$

(note, we have made use of $e^{-T}e^T = 1$, the identity operator, and we are also assuming that we are using orthogonal orbitals, though this does not necessarily have to be true, e.g., valence bond orbitals, and in such cases the last set of equations are not necessarily equal to zero) the latter being the equations to be solved and the former the equation for the evaluation of the energy.

Considering the basic CCSD method:

$$\langle \Phi_0 | e^{-(T_1+T_2)} He^{(T_1+T_2)} | \Phi_0 \rangle = E,$$

$$\langle \Phi_i^a | e^{-(T_1+T_2)} He^{(T_1+T_2)} | \Phi_0 \rangle = 0,$$

$$\langle \Phi_{ij}^{ab} | e^{-(T_1+T_2)} He^{(T_1+T_2)} | \Phi_0 \rangle = 0,$$

in which the similarity transformed Hamiltonian, $\bar{H}$, can be explicitly written down using Hadamard's formula in Lie algebra, also called Hadamard's lemma, though note they are different, in that Hadamard's formula is a lemma of the BCH formula):

$$\bar{H} = e^{-T} He^T = H + [H,T] + \frac{1}{2!}[[H,T],T] + ... = (He^T)_C.$$

The subscript C designates the connected part of the corresponding operator expression.

The resulting similarity transformed Hamiltonian is non-Hermitian, resulting in different left- and right-handed vectors (wave functions) for the same state of interest (this is what is often referred to in coupled cluster theory as the biorthogonality of the solution, or wave function, though it also applies to other non-Hermitian theories as well). The resulting equations are a set of non-linear equations which are solved in an iterative manner. Standard quantum chemistry packages (GAMESS (US), NWChem, ACES II, etc.) solve the coupled cluster equations using the Jacobi method and direct inversion of the iterative subspace (DIIS) extrapolation of the t-amplitudes to accelerate convergence.

Types of Coupled-cluster Methods

The classification of traditional coupled-cluster methods rests on the highest number of excitations allowed in the definition of T. The abbreviations for coupled-cluster methods usually begin with the letters "CC" (for coupled cluster) followed by:

1. S – for single excitations (shortened to *singles* in coupled-cluster terminology)

2. D – for double excitations (*doubles*)

3. T – for triple excitations (*triples*)

4. Q – for quadruple excitations (*quadruples*)

Thus, the T operator in CCSDT has the form

$$T = T_1 + T_2 + T_3.$$

Terms in round brackets indicate that these terms are calculated based on perturbation theory. For example, the CCSD(T) method means:

1. Coupled cluster with a full treatment singles and doubles.

2. An estimate to the connected triples contribution is calculated non-iteratively using Many-Body Perturbation Theory arguments.

General Description of the Theory

The complexity of equations and the corresponding computer codes, as well as the cost of the computation increases sharply with the highest level of excitation. For many applications CCSD, while relatively inexpensive, does not provide sufficient accuracy except for the smallest systems (approximately 2 to 4 electrons), and often an approximate treatment of triples is needed. The most well known coupled cluster method that provides an estimate of connected triples is CCSD(T), which provides a good description of closed-shell molecules near the equilibrium geometry, but breaks down in more complicated situations such as bond breaking and diradicals. Another popular method that makes up for the failings of the standard CCSD(T) approach is CR-CC(2,3), where the triples contribution to the energy is computed from the difference between the exact solution and the CCSD energy, and is not based on perturbation theory arguments. More complicated coupled-cluster methods such as CCSDT and CCSDTQ are used only for high-accuracy calculations of small molecules. The inclusion of all n levels of excitation for the n-electron system gives the exact solution of the Schrödinger equation within the given basis set, within the Born–Oppenheimer approximation (although schemes have also been drawn up to work without the BO approximation).

One possible improvement to the standard coupled-cluster approach is to add terms linear in the interelectronic distances through methods such as CCSD-R12. This improves the treatment of dynamical electron correlation by satisfying the Kato cusp condition and accelerates convergence with respect to the orbital basis set. Unfortunately, R12 methods invoke the resolution of the identity which requires a relatively large basis set in order to be a good approximation.

The coupled-cluster method described above is also known as the *single-reference* (SR) coupled-cluster method because the exponential ansatz involves only one reference function $|\Phi_0\rangle$. The standard generalizations of the SR-CC method are the *multi-reference* (MR) approaches: state-universal coupled cluster (also known as Hilbert space coupled cluster), valence-universal coupled cluster (or Fock space coupled cluster) and state-selective coupled cluster (or state-specific coupled cluster).

Historical Accounts

In the first reference below, Kümmel comments:

> *Considering the fact that the CC method was well understood around the late fifties it looks strange that nothing happened with it until 1966, as Jiří Čížek published his first paper on a quantum chemistry problem. He had looked into the 1957 and 1960 papers published in* Nuclear Physics *by Fritz and myself. I always found it quite remarkable that a quantum chemist would open an issue of a nuclear physics journal. I myself at the time had almost given up the CC method as not tractable and, of course, I never looked into the quantum chemistry journals. The result was that I learnt about Jiří's work as late as in the early seventies, when he sent me a big parcel with reprints of the many papers he and Joe Paldus had written until then.*

Josef Paldus also wrote his first-hand account of the origins of coupled-cluster theory, its implementation, and exploitation in electronic wave function determination; his account is primarily about the making of coupled-cluster theory rather than about the theory itself.

Relation to other Theories

Configuration Interaction

The C_j excitation operators defining the CI expansion of an N-electron system for the wave function $|\Psi_0\rangle$,

$$|\Psi_0\rangle = (1+C)|\Phi_0\rangle,$$

$$C = \sum_{j=1}^{N} C_j,$$

are related to the cluster operators T, since in the limit of including up to T_N in the cluster operator the CC theory must be equal to full CI, we obtain the following relationships

$$C_1 = T_1,$$

$$C_2 = T_2 + \frac{1}{2}(T_1)^2$$

$$C_3 = T_3 + T_1 T_2 + \frac{1}{6}(T_1)^3,$$

$$C_4 = T_4 + \frac{1}{2}(T_2)^2 + T_1 T_3 + \frac{1}{2}(T_1)^2 T_2 + \frac{1}{24}(T_1)^4,$$

etc.

Symmetry Adapted Cluster

The Symmetry adapted cluster (SAC) approach determines the (spin and) symmetry adapted cluster operator

$$S = \sum_I S_I$$

by solving the following system of energy dependent equations,

$$\langle \Phi | (H - E_0) e^S | \Phi \rangle = 0,$$

$$\langle \Phi_{i_1 \ldots i_n}^{a_1 \ldots a_n} | (H - E_0) e^S | \Phi \rangle = 0,$$

$$i_1 < \cdots < i_n, \; a_1 < \cdots < a_n, \; n = 1, \ldots, M_s,$$

where $| \Phi_{i_1 \ldots i_n}^{a_1 \ldots a_n} \rangle$ are the n-tuply excited determinants relative to $| \Phi \rangle$ (usually they are the spin- and symmetry-adapted configuration state functions, in practical implementations), and M_s is the highest-order of excitation included in the SAC operator. If all of the nonlinear terms in e^S are included then the SAC equations become equivalent to the standard coupled-cluster equations of Jiří Čížek. This is due to the cancellation of the energy-dependent terms with the disconnected terms contributing to the product of He^S, resulting in the same set of nonlinear energy-independent equations. Typically, all nonlinear terms, except $\frac{1}{2} S_2^2$ are dropped, as higher-order nonlinear terms are usually small.

Use in Nuclear Physics

In nuclear physics, coupled cluster saw significantly less use than in quantum chemistry during the 1980s and 1990s. More powerful computers as well as advances in theory (such as the inclusion of three-nucleon interactions) have spawned renewed interest in the method since then, and it has been successfully applied to neutron-rich and medium mass nuclei. Coupled cluster is one of several ab initio methods in nuclear physics, and is specifically suitable for nuclei having closed or nearly closed shells.

Quantum Chemistry Composite Methods

Quantum chemistry composite methods (also referred to as thermochemical recipes) are computational chemistry methods that aim for high accuracy by combining the results of several calculations. They combine methods with a high level of theory and a small basis set with methods that employ lower levels of theory with larger basis sets. They are commonly used to calculate thermodynamic quantities such as enthalpies of formation, atomization energies, ionization energies and electron affinities. They aim for chemical accuracy which is usually defined as within 1 kcal/mol of the experimental value. The first systematic model chemistry of this type with broad applicability was called Gaussian-1 (G1) introduced by John Pople. This was quickly replaced by the Gaussian-2 (G2) which has been used extensively. The Gaussian-3 (G3) was introduced later.

Gaussian-n Theories

Gaussian-2 (G2)

The G2 uses seven calculations:

1. the molecular geometry is obtained by a MP2 optimization using the 6-31G(d) basis set and all electrons included in the perturbation. This geometry is used for all subsequent calculations.

2. The highest level of theory is a quadratic configuration interaction calculation with single and double excitations and a triples excitation contribution (QCISD(T)) with the 6-311G(d) basis set. Such a calculation in the Gaussian and Spartan programs also give the MP2 and MP4 energies which are also used.

3. The effect of polarization functions is assessed using an MP4 calculation with the 6-311G(2df,p) basis set.

4. The effect of diffuse functions is assessed using an MP4 calculation with the 6-311+G(d, p) basis set.

5. The largest basis set is 6-311+G(3df,2p) used at the MP2 level of theory.

6. A Hartree–Fock geometry optimization with the 6-31G(d) basis set used to give a geometry for.

7. A frequency calculation with the 6-31G(d) basis set to obtain the zero-point vibrational energy (ZPVE).

The various energy changes are assumed to be additive so the combined energy is given by:

EQCISD(T) from 2 + [EMP4 from 3 - EMP4 from 2] + [EMP4 from 4 - EMP4 from 2] + [EMP2 from 5 + EMP2 from 2 - EMP2 from 3 - EMP2 from 4]

The second term corrects for the effect of adding the polarization functions. The third term corrects for the diffuse functions. The final term corrects for the larger basis set with the terms from steps 2, 3 and 4 preventing contributions from being counted twice. Two final corrections are made to this energy. The ZPVE is scaled by 0.8929. An empirical correction is then added to account for factors not considered above. This is called the higher level correction (HC) and is given by -0.00481 x (number of valence electrons) -0.00019 x (number of unpaired valence electrons). The two numbers are obtained calibrating the results against the experimental results for a set of molecules. The scaled ZPVE and the HLC are added to give the final energy. For some molecules containing one of the third row elements Ga–Xe, a further term is added to account for spin orbit coupling.

Several variants of this procedure have been used. Removing steps 3 and 4 and relying only on the MP2 result from step 5 is significantly cheaper and only slightly less accurate. This is the G2MP2 method. Sometimes the geometry is obtained using a density functional theory method such as B3LYP and sometimes the QCISD(T) method in step 2 is replaced by the coupled cluster method CCSD(T).

The G2(+) variant, where the "+" symbol refers to added diffuse functions, better describes anions than conventional G2 theory. The 6-31+G(d) basis set is used in place of the 6-31G(d) basis set for both the initial geometry optimization, as well as the second geometry optimization and frequency calculation. Additionally, the frozen-core approximation is made for the initial MP2 optimization, whereas G2 usually uses the full calculation.

Gaussian-3 (G3)

The G3 is very similar to G2 but learns from the experience with G2 theory. The 6-311G basis set is replaced by the smaller 6-31G basis. The final MP2 calculations use a larger basis set, generally just called G3large, and correlating all the electrons not just the valence electrons as in G2 theory, additionally a spin-orbit correction term and an empirical correction for valence electrons are introduced. This gives some core correlation contributions to the final energy. The HLC takes the same form but with different empirical parameters.

Gaussian-4 (G4)

Gaussian 4 (G4) theory is an approach for the calculation of energies of molecular species containing first-row (Li–F), second-row (Na–Cl), and third row main group elements. G4 theory is an improved modification of the earlier approach G3 theory. The modifications to G3- theory are the change in an estimate of the Hartree–Fock energy limit, an expanded polarization set for the large basis set calculation, use of CCSD(T) energies, use of geometries from density functional theory and zero-point energies, and two added higher level correction parameters. According to the developers, this theory gives significant improvement over G3-theory.

Feller-Peterson-Dixon Approach (FPD)

Unlike fixed-recipe, "model chemistries", the FPD approach consists of a flexible sequence of (up to) 13 components that vary with the nature of the chemical system under study and the desired accuracy in the final results. In most instances, the primary component relies on coupled cluster theory, such as CCSD(T), or configuration interaction theory combined with large Gaussian basis sets and extrapolation to the complete basis set limit. As with some other approaches, additive corrections for core/valence, scalar relativistic and higher order correlation effects are usually included. Attention is paid to the uncertainties associated with each of the components so as to permit a crude estimate of the uncertainty in the overall results. Accurate structural parameters and vibrational frequencies are a natural byproduct of the method. While the computed molecular properties can be highly accurate, the computationally intensive nature of the FPD approach limits the size of the chemical system to which it can be applied to roughly 10 or fewer first/second row atoms.

The FPD Approach has been heavily benchmarked against experiment. When applied at the highest possible level, FDP is capable to yielding a root-mean-square (RMS) deviation with respect to experiment of 0.29 kcal/mol (287 comparisons covering atomization energies, ionization potentials, electron affinities and proton affinities). In terms of equilibrium, bottom-of-the-well structures, FPD gives an RMS deviation of 0.0020 Å (114 comparisons not involving hydrogens) and 0.0034 Å (54 comparisons involving hydrogen). Similar good agreement was found for vibrational frequencies.

T1

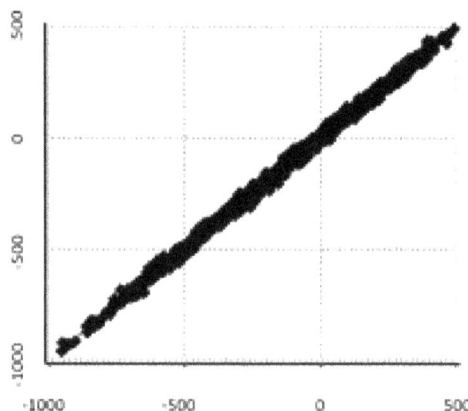

The calculated T1 heat of formation (y axis) compared to the experimental heat of formation (x axis) for a set of >1800 diverse organic molecules from the NIST thermochemical database with mean absolute and RMS errors of 8.5 and 11.5 kJ/mol, respectively.

The T1 method. is an efficient computational approach developed for calculating accurate heats of formation of uncharged, closed-shell molecules comprising H, C, N, O, F, Si, P, S, Cl and Br, within experimental error. It is practical for molecules up to molecular weight ~ 500 a.m.u.

T1 method as incorporated in Spartan consists of:

1. HF/6-31G* optimization.

2. RI-MP2/6-311+G(2d,p)[6-311G*] single point energy with dual basis set.

3. An empirical correction using atom counts, Mulliken bond orders, HF/6-31G* and RI-MP2 energies as variables.

T1 follows the G3(MP2) recipe, however, by substituting an HF/6-31G* for the MP2/6-31G* geometry, eliminating both the HF/6-31G* frequency and QCISD(T)/6-31G* energy and approximating the MP2/G3MP2large energy using dual basis set RI-MP2 techniques, the T1 method reduces computation time by up to 3 orders of magnitude. Atom counts, Mulliken bond orders and HF/6-31G* and RI-MP2 energies are introduced as variables in a linear regression fit to a set of 1126 G3(MP2) heats of formation. The T1 procedure reproduces these values with mean absolute and RMS errors of 1.8 and 2.5 kJ/mol, respectively. T1 reproduces experimental heats of formation for a set of 1805 diverse organic molecules from the NIST thermochemical database with mean absolute and RMS errors of 8.5 and 11.5 kJ/mol, respectively.

Correlation Consistent Composite Approach (ccCA)

This approach, developed at the University of North Texas by Angela K. Wilson's research group, utilizes the correlation consistent basis sets developed by Dunning and co-workers. Unlike the Gaussian-n methods, ccCA does not contain any empirically fitted term. The B3LYP density functional method with the cc-pVTZ basis set, and cc-pV(T+d)Z for third row elements (Na - Ar), are used to determine the equilibrium geometry. Single point calculations are then used to find the reference energy and additional contributions to the energy. The total ccCA energy for main group is calculated by:

$$E_{ccCA} = E_{MP2/CBS} + \Delta E_{CC} + \Delta E_{CV} + \Delta E_{SR} + \Delta E_{ZPE} + \Delta E_{SO}$$

The reference energy $E_{MP2/CBS}$ is the MP2/aug-cc-pVnZ (where n=D,T,Q) energies extrapolated at the complete basis set limit by the Peterson mixed gaussian exponential extrapolation scheme. CCSD(T)/cc-pVTZ is used to account for correlation beyond the MP2 theory:

$$\Delta E_{CC} = E_{CCSD(T)/cc-pVTZ} - E_{MP2/cc-pVTZ}$$

Core-core and core-valence interactions are accounted for using MP2(FC1)/aug-cc-pCVTZ:

$$\Delta E_{CV} = E_{MP2(FC1)/aug-cc-pCVTZ} - E_{MP2/aug-cc-pVTZ}$$

Scalar relativistic effects are also taken into account with a one-particle Douglass Kroll Hess Hamiltonian and recontracted basis sets:

$$\Delta E_{SR} = E_{MP2-DK/cc-pVTZ-DK} - E_{MP2/cc-pVTZ}$$

The last two terms are Zero Point Energy corrections scaled with a factor of 0.989 to account for deficiencies in the harmonic approximation and spin-orbit corrections considered only for atoms.

The Correlation Consistent Composite Approach is available as a keyword in NWChem and GAMESS (ccCA-S4 and ccCA-CC(2,3)).

Complete Basis Set Methods (CBS)

The Complete Basis Set (CBS) methods are a family of composite methods, the members of which are: CBS-4M, CBS-QB3, and CBS-APNO, in increasing order of accuracy. These methods offer errors of 2.5, 1.1, and 0.7 kcal/mol when tested against the G2 test set. The CBS methods were developed by George Petersson and coworkers, and they make extrapolate several single-point energies to the "exact" energy. In comparison, the Gaussian-n methods perform their approximation using additive corrections. Similar to the modified G2(+) method, CBS-QB3 has been modified by the inclusion of diffuse functions in the geometry optimization step to give CBS-QB3(+). The CBS family of methods is available via keywords in the Gaussian 09 suite of programs.

Weizmann-n Theories

The Weizmann-n ab initio methods (Wn, n = 1–4) are highly accurate composite theories devoid of empirical parameters. These theories are capable of sub-kJ/mol accuracies in prediction of fundamental thermochemical quantities such as heats of formation and atomization energies, and unprecedented accuracies in prediction of spectroscopic constants. The ability of these theories to successfully reproduce the CCSD(T)/CBS (W1 and W2), CCSDT(Q)/CBS (W3), and CCSDTQ5/CBS (W4) energies relies on judicious combination of very large Gaussian basis sets with basis-set extrapolation techniques. Thus, the high accuracy of Wn theories comes with the price of a significant computational cost. In practice, for systems consisting of more than ~9 non-hydrogen atoms (with C1 symmetry), even the computationally more economical W1 theory becomes prohibitively expensive with current mainstream server hardware.

In an attempt to extend the applicability of the Wn ab initio thermochemistry methods, explicitly correlated versions of these theories have been developed: Wn-F12 (n = 1–3) and more recently even a W4-F12 theory. W1-F12 was successfully applied to large hydrocarbons (e.g., dodecahedrane, as well as to systems of biological relevance (e.g., DNA bases). W4-F12 theory has been applied to systems as large as benzene.

Density Functional Theory

Density functional theory (DFT) is a computational quantum mechanical modelling method used in physics, chemistry and materials science to investigate the electronic structure (principally the ground state) of many-body systems, in particular atoms, molecules, and the condensed phases. Using this theory, the properties of a many-electron system can be determined by using functionals, i.e. functions of another function, which in this case is the spatially dependent electron density. Hence the name density functional theory comes from the use of functionals of the electron density. DFT is among the most popular and versatile methods available in condensed-matter physics, computational physics, and computational chemistry.

DFT has been very popular for calculations in solid-state physics since the 1970s. However, DFT was not considered accurate enough for calculations in quantum chemistry until the 1990s, when the approximations used in the theory were greatly refined to better model the exchange and correlation interactions. Computational costs are relatively low when compared to traditional methods, such as exchange only Hartree–Fock theory and its descendants that include electron correlation.

Despite recent improvements, there are still difficulties in using density functional theory to properly describe intermolecular interactions (of critical importance to understanding chemical reactions), especially van der Waals forces (dispersion); charge transfer excitations; transition states, global potential energy surfaces, dopant interactions and some other strongly correlated systems; and in calculations of the band gap and ferromagnetism in semiconductors. Its incomplete treatment of dispersion can adversely affect the accuracy of DFT (at least when used alone and uncorrected) in the treatment of systems which are dominated by dispersion (e.g. interacting noble gas atoms) or where dispersion competes significantly with other effects (e.g. in biomolecules). The development of new DFT methods designed to overcome this problem, by alterations to the functional or by the inclusion of additive terms, is a current research topic.

Overview of Method

In the context of computational materials science, ab initio (from first principles) DFT calculations allow the prediction and calculation of material behaviour on the basis of quantum mechanical considerations, without requiring higher order parameters such as fundamental material properties. In contemporary DFT techniques the electronic structure is evaluated using a potential acting on the system's electrons. This DFT potential is constructed as the sum of external potentials (V_{ext}), which is determined solely by the structure and the elemental composition of the system, and an effective potential (V_{eff}), which represents inter-electronic interactions. Thus, a problem for a representative super-cell of a material with n electrons can be studied as a set of n one-electron Schrödinger like equations, which are also known as Kohn–Sham equations.

Although density functional theory has its roots in the Thomas–Fermi model for the electronic structure of materials, DFT was first put on a firm theoretical footing by Walter Kohn and Pierre Hohenberg in the framework of the two Hohenberg–Kohn theorems (H–K). The original H–K theorems held only for non-degenerate ground states in the absence of a magnetic field, although they have since been generalized to encompass these.

The first H–K theorem demonstrates that the ground state properties of a many-electron system are uniquely determined by an electron density that depends on only 3 spatial coordinates. It set down the groundwork for reducing the many-body problem of N electrons with 3N spatial coordinates to 3 spatial coordinates, through the use of functionals of the electron density. This theorem has since been extended to the time-dependent domain to develop time-dependent density functional theory (TDDFT), which can be used to describe excited states.

The second H–K theorem defines an energy functional for the system and proves that the correct ground state electron density minimizes this energy functional.

In work that later won them the Nobel prize in chemistry, The H–K theorem was further developed by Walter Kohn and Lu Jeu Sham to produce Kohn–Sham DFT (KS DFT). Within this framework, the intractable many-body problem of interacting electrons in a static external potential is reduced to a tractable problem of non-interacting electrons moving in an effective potential. The effective potential includes the external potential and the effects of the Coulomb interactions between the electrons, e.g., the exchange and correlation interactions. Modeling the latter two interactions becomes the difficulty within KS DFT. The simplest approximation is the local-density approximation (LDA), which is based upon exact exchange energy for a uniform electron gas, which can be obtained from the Thomas–Fermi model, and from fits to the correlation energy for a uniform electron gas. Non-interacting systems are relatively easy to solve as the wavefunction can be represented as a Slater determinant of orbitals. Further, the kinetic energy functional of such a system is known exactly. The exchange-correlation part of the total-energy functional remains unknown and must be approximated.

Another approach, less popular than KS DFT but arguably more closely related to the spirit of the original H-K theorems, is orbital-free density functional theory (OFDFT), in which approximate functionals are also used for the kinetic energy of the non-interacting system.

Derivation and Formalism

As usual in many-body electronic structure calculations, the nuclei of the treated molecules or clusters are seen as fixed (the Born–Oppenheimer approximation), generating a static external potential V in which the electrons are moving. A stationary electronic state is then described by a wavefunction $\Psi(\vec{r}_1, \ldots, \vec{r}_N)$ satisfying the many-electron time-independent Schrödinger equation

$$\hat{H}\Psi = \left[\hat{T} + \hat{V} + \hat{U}\right]\Psi = \left[\sum_i^N \left(-\frac{\hbar^2}{2m_i}\nabla_i^2\right) + \sum_i^N V(\vec{r}_i) + \sum_{i<j}^N U(\vec{r}_i, \vec{r}_j)\right]\Psi = E\Psi$$

where, for the N-electron system, $\hat{H}$ is the Hamiltonian, E is the total energy, $\hat{T}$ is the kinetic energy, $\hat{V}$ is the potential energy from the external field due to positively charged nuclei, and $\hat{U}$ is the electron-electron interaction energy. The operators $\hat{T}$ and $\hat{U}$ are called universal operators as they are the same for any N-electron system, while $\hat{V}$ is system dependent. This complicated many-particle equation is not separable into simpler single-particle equations because of the interaction term $\hat{U}$.

There are many sophisticated methods for solving the many-body Schrödinger equation based on the expansion of the wavefunction in Slater determinants. While the simplest one is the Hartree–Fock method, more sophisticated approaches are usually categorized as post-Hartree–Fock methods. However, the problem with these methods is the huge computational effort, which makes it virtually impossible to apply them efficiently to larger, more complex systems.

Here DFT provides an appealing alternative, being much more versatile as it provides a way to systematically map the many-body problem, with $\hat{U}$, onto a single-body problem without $\hat{U}$. In DFT the key variable is the electron density $n(\vec{r})$, which for a normalized Ψ is given by

$$n(\vec{r}) = N \int d^3 r_2 \cdots \int d^3 r_N \Psi^*(\vec{r}, \vec{r}_2, \ldots, \vec{r}_N) \Psi(\vec{r}, \vec{r}_2, \ldots, \vec{r}_N).$$

This relation can be reversed, i.e., for a given ground-state density $n_0(\vec{r})$ it is possible, in principle, to calculate the corresponding ground-state wavefunction $\Psi_0(\vec{r}_1, \ldots, \vec{r}_N)$. In other words, Ψ is a unique functional of n_0,

$$\Psi_0 = \Psi[n_0]$$

and consequently the ground-state expectation value of an observable $\hat{O}$ is also a functional of n_0

$$O[n_0] = \left\langle \Psi[n_0] \middle| \hat{O} \middle| \Psi[n_0] \right\rangle.$$

In particular, the ground-state energy is a functional of n_0

$$E_0 = E[n_0] = \left\langle \Psi[n_0] \middle| \hat{T} + \hat{V} + \hat{U} \middle| \Psi[n_0] \right\rangle$$

where the contribution of the external potential $\left\langle \Psi[n_0] \middle| \hat{V} \middle| \Psi[n_0] \right\rangle$ can be written explicitly in terms of the ground-state density n_0

$$V[n_0] = \int V(\vec{r}) n_0(\vec{r}) d^3 r.$$

More generally, the contribution of the external potential $\left\langle \Psi \middle| \hat{V} \middle| \Psi \right\rangle$ can be written explicitly in terms of the density n,

$$V[n] = \int V(\vec{r})n(\vec{r})d^3r.$$

The functionals $T[n]$ and $U[n]$ are called universal functionals, while $V[n]$ is called a non-universal functional, as it depends on the system under study. Having specified a system, i.e., having specified $\hat{V}$, one then has to minimize the functional

$$E[n] = T[n] + U[n] + \int V(\vec{r})n(\vec{r})d^3r$$

with respect to $n(\vec{r})$, assuming one has got reliable expressions for $T[n]$ and $U[n]$. A successful minimization of the energy functional will yield the ground-state density n_0 and thus all other ground-state observables.

The variational problems of minimizing the energy functional $E[n]$ can be solved by applying the Lagrangian method of undetermined multipliers. First, one considers an energy functional that doesn't explicitly have an electron-electron interaction energy term,

$$E_s[n] = \left\langle \Psi_s[n] \middle| \hat{T} + \hat{V}_s \middle| \Psi_s[n] \right\rangle$$

where $\hat{T}$ denotes the kinetic energy operator and $\hat{V}_s$ is an external effective potential in which the particles are moving, so that $n_s(\vec{r}) \overset{\text{def}}{=} n(\vec{r})$.

Thus, one can solve the so-called Kohn–Sham equations of this auxiliary non-interacting system,

$$\left[-\frac{\hbar^2}{2m}\nabla^2 + V_s(\vec{r}) \right]\phi_i(\vec{r}) = \epsilon_i\phi_i(\vec{r})$$

which yields the orbitals ϕ_i that reproduce the density $n(\vec{r})$ of the original many-body system

$$n(\vec{r}) \overset{\text{def}}{=} n_s(\vec{r}) = \sum_i^N |\phi_i(\vec{r})|^2.$$

The effective single-particle potential can be written in more detail as

$$V_s(\vec{r}) = V(\vec{r}) + \int \frac{e^2 n_s(\vec{r}')}{|\vec{r} - \vec{r}'|}d^3r' + V_{XC}[n_s(\vec{r})]$$

where the second term denotes the so-called Hartree term describing the electron-electron Coulomb repulsion, while the last term V_{XC} is called the exchange-correlation potential. Here, V_{XC} includes all the many-particle interactions. Since the Hartree term and V_{XC} depend on $n(\vec{r})$, which depends on the ϕ_i, which in turn depend on V_s, the problem of solving the Kohn–Sham equation has to be done in a self-consistent (i.e., iterative) way. Usually one starts with an initial guess for $n(\vec{r})$, then calculates the corresponding V_s and solves the Kohn–Sham equations for the ϕ_i, From these one calculates a new density and starts again. This procedure is then repeated until convergence is reached. A non-iterative approximate formulation called Harris functional DFT is an alternative approach to this.

NOTE 1: The one-to-one correspondence between electron density and single-particle potential is not so smooth. It contains kinds of non-analytic structure. $E_s[n]$ contains kinds of singularities, cuts and branches. This may indicate a limitation of our hope for representing exchange-correlation functional in a simple analytic form.

NOTE 2: It is possible to extend the DFT idea to the case of Green function G instead of the density n. It is called as Luttinger–Ward functional (or kinds of similar functionals), written as $E[G]$. However, G is determined not as its minimum, but as its extremum. Thus we may have some theoretical and practical difficulties.

NOTE 3: There is no one-to-one correspondence between one-body density matrix $n(\vec{r}, \vec{r}')$ and the one-body potential $V(\vec{r}, \vec{r}')$. (Remember that all the eigenvalues of $n(\vec{r}, \vec{r}')$ is unity). In other words, it ends up with a theory similar as the Hartree-Fock (or hybrid) theory.

Relativistic Density Functional Theory (Explicit Functional Forms)

The same theorems can be proven in the case of relativistic electrons thereby providing generalization of DFT for the relativistic case. Unlike nonrelativistic theory, in the relativistic case it is possible to derive a few exact and explicit formulas for relativistic density functional.

Let one consider an electron in a hydrogen-like ion obeying the relativistic Dirac equation. Hamiltonian H for relativistic electron moving in the Coulomb potential can be chosen in the following form (atomic units are used):

$$H = c(\vec{\alpha} \cdot \vec{p}) + eV + mc^2 \beta,$$

where $V = -\dfrac{eZ}{r}$ is Coulomb potential of a point-like nucleus, $\vec{p}$ is a momentum operator of electron, e, m and c are electron electric charge, mass and speed of light in vacuum constants respectively, and finally $\vec{\alpha}$ and β are set of Dirac 4×4 matrixes:

$$\vec{\alpha} = \begin{pmatrix} 0 & \vec{\sigma} \\ \vec{\sigma} & 0 \end{pmatrix}, \quad \beta = \begin{pmatrix} I & 0 \\ 0 & -I \end{pmatrix}.$$

To find out eigen functions and corresponding energies one solves the eigen function equation:

$$H\Psi = E\Psi,$$

where $\Psi = \left(\Psi(1), \Psi(2), \Psi(3), \Psi(4)\right)^T$ is a four component wave function and E is associated eigen energy. It is demonstrated in the section that application of the virial theorem to eigenfunction equation produces the following formula for eigen energy of any bound state:

$$E = mc^2 \left\langle \Psi \left| \beta \right| \Psi \right\rangle = mc^2 \int |\Psi(1)|^2 + |\Psi(2)|^2 - |\Psi(3)|^2 - |\Psi(4)|^2 \, d\tau,$$

and analogously the virial theorem applied to the eigenfunction equation with squared Hamiltonian yields:

$$E^2 = m^2 c^4 + emc^2 \left\langle \Psi \left| V\beta \right| \Psi \right\rangle.$$

It is easy to see that both written above formulas represent density functionals. The former formula can be easily generalized for multi-electron case.

Approximations (Exchange-correlation Functionals)

The major problem with DFT is that the exact functionals for exchange and correlation are not known except for the free electron gas. However, approximations exist which permit the calculation of certain physical quantities quite accurately. In physics the most widely used approximation is the local-density approximation (LDA), where the functional depends only on the density at the coordinate where the functional is evaluated:

$$E_{XC}^{LDA}[n] \quad \int \epsilon_{XC}(n)n(r)\mathrm{d}^3 r.$$

The local spin-density approximation (LSDA) is a straightforward generalization of the LDA to include electron spin:

$$E_{XC}^{LSDA}\left[n_\uparrow, n_\downarrow\right] = \int \epsilon_{XC}\left(n_\uparrow, n_\downarrow\right)n(\vec{r})\mathrm{d}^3 r.$$

Highly accurate formulae for the exchange-correlation energy density $\epsilon_{XC}(n_\uparrow, n_\downarrow)$ have been constructed from quantum Monte Carlo simulations of jellium.

The LDA assumes that the density is the same everywhere. Because of this, the LDA has a tendency to under-estimate the exchange energy and over-estimate the correlation energy. The errors due to the exchange and correlation parts tend to compensate each other to a certain degree. To correct for this tendency, it is common to expand in terms of the gradient of the density in order to account for the non-homogeneity of the true electron density. This allows for corrections based on the changes in density away from the coordinate. These expansions are referred to as generalized gradient approxima-tions (GGA) and have the following form:

$$E_{XC}^{GGA}[n_\uparrow, n_\downarrow] = \int \epsilon_{XC}\left(n_\uparrow, n_\downarrow, \vec{\nabla} n_\uparrow, \vec{\nabla} n_\downarrow\right) n(\vec{r}) d^3r.$$

Using the latter (GGA), very good results for molecular geometries and ground-state energies have been achieved.

Potentially more accurate than the GGA functionals are the meta-GGA functionals, a natural development after the GGA (generalized gradient approximation). Meta-GGA DFT functional in its original form includes the second derivative of the electron densi-ty (the Laplacian) whereas GGA includes only the density and its first derivative in the exchange-correlation potential.

Functionals of this type are, for example, TPSS and the Minnesota Functionals. These functionals include a further term in the expansion, depending on the density, the gra-dient of the density and the Laplacian (second derivative) of the density.

Difficulties in expressing the exchange part of the energy can be relieved by including a component of the exact exchange energy calculated from Hartree–Fock theory. Func-tionals of this type are known as hybrid functionals.

Generalizations to Include Magnetic Fields

The DFT formalism described above breaks down, to various degrees, in the presence of a vector potential, i.e. a magnetic field. In such a situation, the one-to-one mapping between the ground-state electron density and wavefunction is lost. Generalizations to include the effects of magnetic fields have led to two different theories: current density functional theory (CDFT) and magnetic field density functional theory (BDFT). In both these theo-ries, the functional used for the exchange and correlation must be generalized to include more than just the electron density. In current density functional theory, developed by Vignale and Rasolt, the functionals become dependent on both the electron density and the paramagnetic current density. In magnetic field density functional theory, developed by Salsbury, Grayce and Harris, the functionals depend on the electron density and the magnetic field, and the functional form can depend on the form of the magnetic field. In both of these theories it has been difficult to develop functionals beyond their equivalent to LDA, which are also readily implementable computationally. Recently an extension by Pan and Sahni extended the Hohenberg-Kohn theorem for non constant magnetic fields using the density and the current density as fundamental variables.

Applications

In general, density functional theory finds increasingly broad application in the chem-ical and material sciences for the interpretation and prediction of complex system behavior at an atomic scale. Specifically, DFT computational methods are applied for the study of systems to synthesis and processing parameters. In such systems, experimental studies are often encumbered by inconsistent results and non-equilibrium conditions. Examples of contemporary DFT applications include studying the effects of dopants on phase transformation behavior in oxides, magnetic behaviour in dilute magnetic semiconductor materials and the study of magnetic and electronic behavior in ferroelectrics and dilute magnetic semiconductors.. Also, it has been shown that DFT has a good results in the prediction of sensitivity of some nanostructures to environment pollutants like SO_2 or Acrolein as well as prediction of mechanical properties.

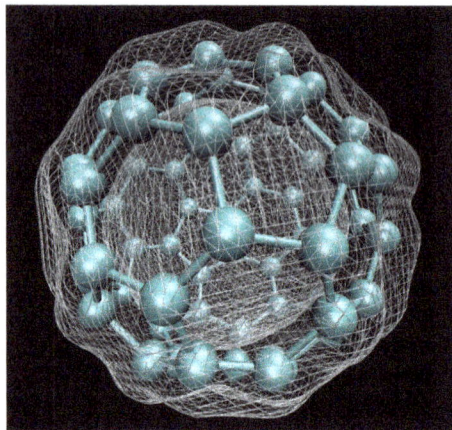

C_{60} with isosurface of ground-state electron density as calculated with DFT

In practice, Kohn–Sham theory can be applied in several distinct ways depending on what is being investigated. In solid state calculations, the local density approximations are still commonly used along with plane wave basis sets, as an electron gas approach is more appropriate for electrons delocalised through an infinite solid. In molecular calculations, however, more sophisticated functionals are needed, and a huge variety of exchange-correlation functionals have been developed for chemical applications. Some of these are inconsistent with the uniform electron gas approximation, however, they must reduce to LDA in the electron gas limit. Among physicists, probably the most widely used functional is the revised Perdew–Burke–Ernzerhof exchange model (a direct generalized-gradient parametrization of the free electron gas with no free parameters); however, this is not sufficiently calorimetrically accurate for gas-phase molecular calculations. In the chemistry community, one popular functional is known as BLYP (from the name Becke for the exchange part and Lee, Yang and Parr for the correlation part). Even more widely used is B3LYP which is a hybrid functional in which the exchange energy, in this case from Becke's exchange functional, is combined with the exact energy from Hartree–Fock theory. Along with the component exchange and correlation functionals, three parameters define the hybrid functional, specifying how much of the exact exchange is mixed in. The adjustable parameters in hybrid functionals are generally fitted to a 'training set' of molecules. Unfortunately, although the results obtained with these functionals are usually sufficiently accurate for most applications, there is no systematic way of improving them (in contrast to some of the traditional wavefunction-based methods like configuration interaction or coupled cluster theory). Hence in the current DFT approach it is not possible to estimate the error of the calculations without comparing them to other methods or experiments.

Thomas–Fermi Model

The predecessor to density functional theory was the Thomas–Fermi model, developed independently by both Thomas and Fermi in 1927. They used a statistical model to approximate the distribution of electrons in an atom. The mathematical basis postulated that electrons are distributed uniformly in phase space with two electrons in every h^3 of volume. For each element of coordinate space volume d^3r we can fill out a sphere of momentum space up to the Fermi momentum p_f

$$\frac{4}{3}\pi p_f^3(\vec{r}).$$

Equating the number of electrons in coordinate space to that in phase space gives:

$$n(\vec{r}) = \frac{8\pi}{3h^3} p_f^3(\vec{r}).$$

Solving for p_f and substituting into the classical kinetic energy formula then leads directly to a kinetic energy represented as a functional of the electron density:

$$t_{TF}[n] = \frac{p^2}{2m_e} \propto \frac{(n^{\frac{1}{3}})^2}{2m_e} \propto n^{\frac{2}{3}}(\vec{r})$$

$$T_{TF}[n] = C_F \int n(\vec{r}) n^{\frac{2}{3}}(\vec{r}) d^3r = C_F \int n^{\frac{5}{3}}(\vec{r}) d^3r$$

where $C_F = \frac{3h^2}{10m_e}\left(\frac{3}{8\pi}\right)^{\frac{2}{3}}$.

As such, they were able to calculate the energy of an atom using this kinetic energy functional combined with the classical expressions for the nuclear-electron and electron-electron interactions (which can both also be represented in terms of the electron density).

Although this was an important first step, the Thomas–Fermi equation's accuracy is limited because the resulting kinetic energy functional is only approximate, and because the method does not attempt to represent the exchange energy of an atom as a conclusion of the Pauli principle. An exchange energy functional was added by Dirac in 1928.

However, the Thomas–Fermi–Dirac theory remained rather inaccurate for most applications. The largest source of error was in the representation of the kinetic energy, followed by the errors in the exchange energy, and due to the complete neglect of electron correlation.

Teller (1962) showed that Thomas–Fermi theory cannot describe molecular bonding. This can be overcome by improving the kinetic energy functional.

The kinetic energy functional can be improved by adding the Weizsäcker (1935) correction:

$$T_W[n] = \frac{\hbar^2}{8m} \int \frac{|\nabla n(\vec{r})|^2}{n(\vec{r})} d^3r.$$

Hohenberg–Kohn Theorems

The Hohenberg-Kohn theorems relate to any system consisting of electrons moving under the influence of an external potential.

Theorem 1. The external potential (and hence the total energy), is a unique functional of the electron density.

If two systems of electrons, one trapped in a potential $v_1(\vec{r})$ and the other in $v_2(\vec{r})$, have the same ground-state density $n(\vec{r})$ then necessarily $v_1(\vec{r}) - v_2(\vec{r}) = const.$

Corollary: the ground state density uniquely determines the potential and thus all properties of the system, including the many-body wave function. In particular, the "HK" functional, defined as $F[n] = T[n] + U[n]$ is a universal functional of the density (not depending explicitly on the external potential).

Theorem 2. The functional that delivers the ground state energy of the system, gives the lowest energy if and only if the input density is the true ground state density.

For any positive integer N and potential $v(\vec{r})$, a density functional $F[n]$ exists such that $E_{(v,N)}[n] = F[n] + \int v(\vec{r}) n(\vec{r}) d^3r$ obtains its minimal value at the ground-state density of N electrons in the potential $v(\vec{r})$. The minimal value of $E_{(v,N)}[n]$ is then the ground state energy of this system.

Pseudo-potentials

The many electron Schrödinger equation can be very much simplified if electrons are divided in two groups: valence electrons and inner core electrons. The electrons in the inner shells are strongly bound and do not play a significant role in the chemical binding of atoms; they also partially screen the nucleus, thus forming with the nucleus an almost inert core. Binding properties are almost completely due to the valence electrons, especially in metals and semiconductors. This separation suggests that inner electrons can be ignored in a large number of cases, thereby reducing the atom to an ionic core that interacts with the valence electrons. The use of an effective interaction, a pseudopotential, that approximates the potential felt by the valence electrons, was first proposed by Fermi in 1934 and Hellmann in 1935. In spite of the simplification pseudo-potentials introduce in calculations, they remained forgotten until the late 50's.

Ab Initio Pseudo-potentials

A crucial step toward more realistic pseudo-potentials was given by Topp and Hopfield and more recently Cronin, who suggested that the pseudo-potential should be adjusted such that they describe the valence charge density accurately. Based on that idea, modern pseudo-potentials are obtained inverting the free atom Schrödinger equation for a given reference electronic configuration and forcing the pseudo wave-functions to coincide with the true valence wave functions beyond a certain distance rl. The pseudo wave-functions are also forced to have the same norm as the true valence wave-functions and can be written as

$$R_l^{pp}(r) = R_{nl}^{AE}(r).$$

$$\int_0^{rl} dr \mid R_l^{PP}(r) \mid^2 r^2 = \int_0^{rl} dr \mid R_{nl}^{AE}(r) \mid^2 r^2.$$

where $R_l(r)$. is the radial part of the wavefunction with angular momentum l, and pp and AE denote, respectively, the pseudo wave-function and the true (all-electron) wave-function. The index n in the true wave-functions denotes the valence level. The distance beyond which the true and the pseudo wave-functions are equal, rl, is also l-dependent.

Electron Smearing

The electrons of system will occupy the lowest Kohn-Sham eigenstates up to a given energy level according to the Aufbau principle. This corresponds to the step-like Fermi-Dirac distribution at absolute zero. If there are several degenerate or close to degenerate eigenstates at the Fermi level, it is possible to get convergence problems, since very small perturbations may change the electron occupation. One way of damping these oscillations is to *smear* the electrons, i.e. allowing fractional occupancies. One approach of doing this is to assign a finite temperature to the electron Fermi-Dirac distribution. Other ways is to assign a cumulative Gaussian distribution of the electrons or using a Methfessel-Paxton method.

Software Supporting DFT

DFT is supported by many Quantum chemistry and solid state physics software packages, often along with other methods.

Semi-empirical Quantum Chemistry Method

Semi-empirical quantum chemistry methods are based on the Hartree–Fock formalism, but make many approximations and obtain some parameters from empirical data. They are very important in computational chemistry for treating large molecules where the full Hartree–Fock method without the approximations is too expensive. The use of empirical parameters appears to allow some inclusion of electron correlation effects into the methods.

Within the framework of Hartree–Fock calculations, some pieces of information (such as two-electron integrals) are sometimes approximated or completely omitted. In order to correct for this loss, semi-empirical methods are parametrized, that is their results are fitted by a set of parameters, normally in such a way as to produce results that best agree with experimental data, but sometimes to agree with *ab initio* results.

Type of Simplifications used

Semi-empirical methods follow what are often called empirical methods where the two-electron part of the Hamiltonian is not explicitly included. For π-electron systems, this was the Hückel method proposed by Erich Hückel. For all valence electron systems, the extended Hückel method was proposed by Roald Hoffmann.

Semi-empirical calculations are much faster than their *ab initio* counterparts, mostly due to the use of the zero differential overlap approximation. Their results, however, can be very wrong if the molecule being computed is not similar enough to the molecules in the database used to parametrize the method.

Preferred Application Domains

Semi-empirical calculations have been most successful in the description of organic chemistry, where only a few elements are used extensively and molecules are of moderate size. However, semi-empirical methods were also applied to solids and nanostructures but with different parameterization.

Empirical research is a way of gaining knowledge by means of direct and indirect observation or experience. As with empirical methods, we can distinguish methods that are:

Methods Restricted to π-electrons

These methods exist for the calculation of electronically excited states of polyenes, both cyclic and linear. These methods, such as the Pariser–Parr–Pople method (PPP), can provide good estimates of the π-electronic excited states, when parameterized well. Indeed, for many years, the PPP method outperformed ab initio excited state calculations.

Methods Restricted to all Valence Electrons

These methods can be grouped into several groups:

- Methods such as CNDO/2, INDO and NDDO that were introduced by John Pople. The implementations aimed to fit, not experiment, but ab initio minimum basis set results. These methods are now rarely used but the methodology is often the basis of later methods.

- Methods that are in the MOPAC, AMPAC, and/or SPARTAN computer programs originally from the group of Michael Dewar. These are MINDO, MNDO, AM1, PM3, RM1 , PM6 and SAM1. Here the objective is to use parameters to fit experimental heats of formation, dipole moments, ionization potentials, and geometries.

- Methods whose primary aim is to predict the geometries of coordination compounds, such as Sparkle/AM1, available for lanthanide complexes.

- Methods whose primary aim is to calculate excited states and hence predict electronic spectra. These include ZINDO and SINDO.

the latter being by far the largest group of methods.

Extended Hückel Method

The extended Hückel method is a semiempirical quantum chemistry method, developed by Roald Hoffmann since 1963. It is based on the Hückel method but, while the original Hückel method only considers pi orbitals, the extended method also includes the sigma orbitals.

The extended Hückel method can be used for determining the molecular orbitals, but it is not very successful in determining the structural geometry of an organic molecule. It can however determine the relative energy of different geometrical configurations. It involves calculations of the electronic interactions in a rather simple way for which the electron-electron repulsions are not explicitly included and the total energy is just a sum of terms for each electron in the molecule. The off-diagonal Hamiltonian matrix elements are given by an approximation due to Wolfsberg and Helmholz that relates them to the diagonal elements and the overlap matrix element.

$$H_{ij} = KS_{ij} \frac{H_{ii} + H_{jj}}{2}$$

K is the Wolfsberg-Helmholtz constant, and is usually given a value of 1.75. In the extended Hückel method, only valence electrons are considered; the core electron energies and functions are supposed to be more or less constant between atoms of the same type. The method uses a series of parametrized energies calculated from atomic ionization potentials or theoretical methods to fill the diagonal of the Fock matrix. After filling the non-diagonal elements and diagonalizing the resulting Fock matrix, the energies (eigenvalues) and wavefunctions (eigenvectors) of the valence orbitals are found.

It is common in many theoretical studies to use the extended Hückel molecular orbitals as a preliminary step to determining the molecular orbitals by a more sophisticated method such as the CNDO/2 method and ab initio quantum chemistry methods. Since the extended Hückel basis set is fixed, the monoparticle calculated wavefunctions must be projected to the basis set where the accurate calculation is to be done. One usually does this by adjusting the orbitals in the new basis to the old ones by least squares method. As only valence electron wavefunctions are found by this method, one must fill the core electron functions by orthonormalizing the rest of the basis set with the calculated orbitals and then selecting the ones with less energy. This leads to the determination of more accurate structures and electronic properties, or in the case of ab initio methods, to somewhat faster convergence.

The method was first used by Roald Hoffmann who developed, with Robert Burns Woodward, rules for elucidating reaction mechanisms (the Woodward–Hoffmann rules). He used pictures of the molecular orbitals from extended Hückel theory to work out the orbital interactions in these cycloaddition reactions.

A closely similar method was used earlier by Hoffmann and William Lipscomb for studies of boron hydrides. The off-diagonal Hamiltonian matrix elements were given as proportional to the overlap integral.

$$H_{ij} = K S_{ij}$$

This simplification of the Wolfsberg and Helmholz approximation is reasonable for boron hydrides as the diagonal elements are reasonably similar due to the small difference in electronegativity between boron and hydrogen.

The method works poorly for molecules that contain atoms of very different electronegativity. To overcome this weakness, several groups have suggested iterative schemes that depend on the atomic charge. One such method, that is still widely used in inorganic and organometallic chemistry is the Fenske-Hall method.

A program for the *extended Hückel method* is YAeHMOP which stands for "yet another extended Hückel molecular orbital package".

Hartree–Fock Method

In computational physics and chemistry, the Hartree–Fock (HF) method is a method of approximation for the determination of the wave function and the energy of a quantum many-body system in a stationary state.

The Hartree–Fock method often assumes that the exact, N-body wave function of the system can be approximated by a single Slater determinant (in the case where the particles are fermions) or by a single permanent (in the case of bosons) of N spin-orbitals. By invoking the variational method, one can derive a set of N-coupled equations for the N spin orbitals. A solution of these equations yields the Hartree–Fock wave function and energy of the system.

Especially in the older literature, the Hartree–Fock method is also called the self-consistent field method (SCF). In deriving what is now called the Hartree equation as an approximate solution of the Schrödinger equation, Hartree required the final field as computed from the charge distribution to be "self-consistent" with the assumed initial field. Thus, self-consistency was a requirement of the solution. The solutions to the non-linear Hartree–Fock equations also behave as if each particle is subjected to the mean field created by all other particles and hence the terminology continued. The equations are almost universally solved by means of an iterative method, although the fixed-point iteration algorithm does not always converge. This solution scheme is not the only one possible and is not an essential feature of the Hartree–Fock method.

The Hartree–Fock method finds its typical application in the solution of the Schrödinger equation for atoms, molecules, nanostructures and solids but it has also found widespread use in nuclear physics. In atomic structure theory, calculations may be for a spectrum with many excited energy levels and consequently the Hartree–Fock method for atoms assumes the wave function is a single configuration state function with well-defined quantum numbers and that the energy level is not necessarily the ground state.

For both atoms and molecules, the Hartree–Fock solution is the central starting point for most methods that describe the many-electron system more accurately.

The rest will focus on applications in electronic structure theory suitable for molecules with the atom as a special case. The discussion here is only for the Restricted Hartree–Fock method, where the atom or molecule is a closed-shell system with all orbitals (atomic or molecular) doubly occupied. Open-shell systems, where some of the electrons are not paired, can be dealt with by one of two Hartree–Fock methods:

- Restricted open-shell Hartree–Fock (ROHF)

- Unrestricted Hartree–Fock (UHF)

Brief History

The origin of the Hartree–Fock method dates back to the end of the 1920s, soon after the discovery of the Schrödinger equation in 1926. In 1927 D. R. Hartree introduced a procedure, which he called the self-consistent field method, to calculate approximate wave functions and energies for atoms and ions. Hartree was guided by some earlier, semi-empirical methods of the early 1920s (by E. Fues, R. B. Lindsay, and himself) set in the old quantum theory of Bohr.

In the Bohr model of the atom, the energy of a state with principal quantum number n is given in atomic units as $E = -1/n^2$. It was observed from atomic spectra that the energy levels of many-electron atoms are well described by applying a modified version of Bohr's formula. By introducing the quantum defect d as an empirical parameter, the energy levels of a generic atom were well approximated by the formula $E = -1/(n+d)^2$, in the sense that one could reproduce fairly well the observed transitions levels observed in the X-ray region (for example, empirical discussion and derivation in Moseley's law). The existence of a non-zero quantum defect was attributed to electron-electron repulsion, which clearly does not exist in the isolated hydrogen atom. This repulsion resulted in partial screening of the bare nuclear charge. These early researchers later introduced other potentials containing additional empirical parameters with the hope of better reproducing the experimental data.

Hartree sought to do away with empirical parameters and solve the many-body time-independent Schrödinger equation from fundamental physical principles, i.e., ab initio. His first proposed method of solution became known as the Hartree method. However, many of Hartree's contemporaries did not understand the physical reasoning behind the Hartree method: it appeared to many people to contain empirical elements, and its connection to the solution of the many-body Schrödinger equation was unclear. However, in 1928 J. C. Slater and J. A. Gaunt independently showed that the Hartree method could be couched on a sounder theoretical basis by applying the variational principle to an ansatz (trial wave function) as a product of single-particle functions.

In 1930 Slater and V. A. Fock independently pointed out that the Hartree method did not respect the principle of antisymmetry of the wave function. The Hartree method used the Pauli exclusion principle in its older formulation, forbidding the presence of two electrons in the same quantum state. However, this was shown to be fundamentally incomplete in its neglect of quantum statistics.

It was then shown that a Slater determinant, a determinant of one-particle orbitals first used by Heisenberg and Dirac in 1926, trivially satisfies the antisymmetric property of the exact solution and hence is a suitable ansatz for applying the variational principle. The original Hartree method can then be viewed as an approximation to the Hartree–Fock method by neglecting exchange. Fock's original method relied heavily on group theory and was too abstract for contemporary physicists to understand and implement. In 1935 Hartree reformulated the method more suitably for the purposes of calculation.

The Hartree–Fock method, despite its physically more accurate picture, was little used until the advent of electronic computers in the 1950s due to the much greater computational demands over the early Hartree method and empirical models. Initially, both the Hartree method and the Hartree–Fock method were applied exclusively to atoms, where the spherical symmetry of the system allowed one to greatly simplify the problem. These approximate methods were (and are) often used together with the central field approximation, to impose that electrons in the same shell have the same radial part, and to restrict the variational solution to be a spin eigenfunction. Even so, solution by hand of the Hartree–Fock equations for a medium-sized atom were laborious; small molecules required computational resources far beyond what was available before 1950.

Hartree–Fock Algorithm

The Hartree–Fock method is typically used to solve the time-independent Schrödinger equation for a multi-electron atom or molecule as described in the Born–Oppenheimer approximation. Since there are no known solutions for many-electron systems (there **are** solutions for one-electron systems such as hydrogenic atoms and the diatomic hydrogen cation), the problem is solved numerically. Due to the nonlinearities introduced by the Hartree–Fock approximation, the equations are solved using a nonlinear method such as iteration, which gives rise to the name "self-consistent field method."

Approximations

The Hartree–Fock method makes five major simplifications in order to deal with this task:

- The Born–Oppenheimer approximation is inherently assumed. The full molecular wave function is actually a function of the coordinates of each of the nuclei, in addition to those of the electrons.

- Typically, relativistic effects are completely neglected. The momentum operator is assumed to be completely non-relativistic.

- The variational solution is assumed to be a linear combination of a finite number of basis functions, which are usually (but not always) chosen to be orthogonal. The finite basis set is assumed to be approximately complete.

- Each energy eigenfunction is assumed to be describable by a single Slater determinant, an antisymmetrized product of one-electron wave functions (i.e., orbitals).

- The mean field approximation is implied. Effects arising from deviations from this assumption are neglected. These effects are often collectively used as a definition of the term electron correlation. However, the label "electron correlation" strictly spoken encompasses both Coulomb correlation and Fermi correlation, and the latter is an effect of electron exchange, which is fully accounted for in the Hartree–Fock method. Stated in this terminology, the method only neglects the Coulomb correlation. However, this is an important flaw, accounting for (among others) Hartree-Fock's inability to capture London dispersion.

Relaxation of the last two approximations give rise to many so-called post-Hartree–Fock methods.

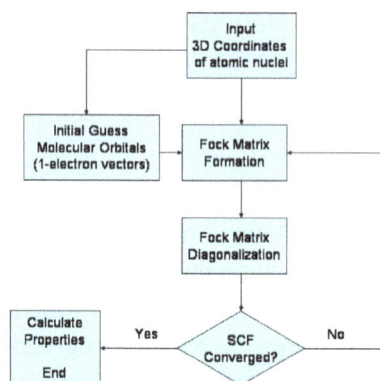

Algorithmic flowchart illustrating the Hartree–Fock method

Variational Optimization of Orbitals

The variational theorem states that for a time-independent Hamiltonian operator, any trial wave function will have an energy expectation value that is greater than or equal to the true ground state wave function corresponding to the given Hamiltonian. Because of this, the Hartree–Fock energy is an upper bound to the true ground state energy of a given molecule. In the context of the Hartree–Fock method, the best possible solution is at the *Hartree–Fock limit*; i.e., the limit of the Hartree–Fock energy as the basis set approaches completeness. (The other is the *full-CI limit*, where the last two approximations of the Hartree–Fock theory as described above are completely undone. It is only when both limits are attained that the exact solution, up to the Born–Oppenheimer approximation, is obtained.) The Hartree–Fock energy is the minimal energy for a single Slater determinant.

The starting point for the Hartree–Fock method is a set of approximate one-electron wave functions known as *spin-orbitals*. For an atomic orbital calculation, these are typically the orbitals for a hydrogenic atom (an atom with only one electron, but the appropriate nuclear charge). For a molecular orbital or crystalline calculation, the initial approximate one-electron wave functions are typically a linear combination of atomic orbitals (LCAO).

The orbitals above only account for the presence of other electrons in an average manner. In the Hartree–Fock method, the effect of other electrons are accounted for in a mean-field theory context. The orbitals are optimized by requiring them to minimize the energy of the respective Slater determinant. The resultant variational conditions on the orbitals lead to a new one-electron operator, the Fock operator. At the minimum, the occupied orbitals are eigensolutions to the Fock operator via a unitary transformation between themselves. The Fock operator is an effective one-electron Hamiltonian operator being the sum of two terms. The first is a sum of kinetic energy operators for each electron, the internuclear repulsion energy, and a sum of nuclear-electronic Coulombic attraction terms. The second are Coulombic repulsion terms between electrons in a mean-field theory description; a net repulsion energy for each electron in the system, which is calculated by treating all of the other electrons within the molecule as a smooth distribution of negative charge. This is the major simplification inherent in the Hartree–Fock method, and is equivalent to the fifth simplification in the above list.

Since the Fock operator depends on the orbitals used to construct the corresponding Fock matrix, the eigenfunctions of the Fock operator are in turn new orbitals which can be used to construct a new Fock operator. In this way, the Hartree–Fock orbitals are optimized iteratively until the change in total electronic energy falls below a predefined threshold. In this way, a set of self-consistent one-electron orbitals are calculated. The Hartree–Fock electronic wave function is then the Slater determinant constructed out of these orbitals. Following the basic postulates of quantum mechanics, the Hartree–Fock wave function can then be used to compute any desired chemical or physical property within the framework of the Hartree–Fock method and the approximations employed.

Mathematical Formulation

The Fock Operator

Because the electron-electron repulsion term of the molecular Hamiltonian involves the coordinates of two different electrons, it is necessary to reformulate it in an approximate way. Under this approximation, (outlined under Hartree–Fock algorithm), all of the terms of the exact Hamiltonian except the nuclear-nuclear repulsion term are re-expressed as the sum of one-electron operators outlined below, for closed-shell atoms or molecules (with two electrons in each spatial orbital). The "(1)" following each operator symbol simply indicates that the operator is 1-electron in nature.

$$\hat{F}[\{\phi_j\}](1) = \hat{H}^{core}(1) + \sum_{j=1}^{N/2}[2\hat{J}_j(1) - \hat{K}_j(1)]$$

where

$$\hat{F}[\{\phi_j\}](1)$$

is the one-electron Fock operator generated by the orbitals ϕ_j, and

$$\hat{H}^{\text{core}}(1) = -\frac{1}{2}\nabla_1^2 - \sum_\alpha \frac{Z_\alpha}{r_{1\alpha}}$$

is the one-electron core Hamiltonian. Also

$$\hat{J}_j(1)$$

is the Coulomb operator, defining the electron-electron repulsion energy due to each of the two electrons in the jth orbital. Finally

$$\hat{K}_j(1)$$

is the exchange operator, defining the electron exchange energy due to the antisymmetry of the total n-electron wave function. This "exchange energy" operator, K, is simply an artifact of the Slater determinant. Finding the Hartree–Fock one-electron wave functions is now equivalent to solving the eigenfunction equation:

$$\hat{F}(1)\phi_i(1) = \epsilon_i \phi_i(1)$$

where $\phi_i(1)$ are a set of one-electron wave functions, called the Hartree–Fock molecular orbitals.

Linear Combination of Atomic Orbitals

Typically, in modern Hartree–Fock calculations, the one-electron wave functions are approximated by a linear combination of atomic orbitals. These atomic orbitals are called Slater-type orbitals. Furthermore, it is very common for the "atomic orbitals" in use to actually be composed of a linear combination of one or more Gaussian-type orbitals, rather than Slater-type orbitals, in the interests of saving large amounts of computation time.

Various basis sets are used in practice, most of which are composed of Gaussian functions. In some applications, an orthogonalization method such as the Gram–Schmidt process is performed in order to produce a set of orthogonal basis functions. This can in principle save computational time when the computer is solving the Roothaan–Hall equations by converting the overlap matrix effectively to an identity matrix. However, in most modern computer programs for molecular Hartree–Fock calculations this procedure is not followed due to the high numerical cost of orthogonalization and the advent of more efficient, often sparse, algorithms for solving the generalized eigenvalue problem, of which the Roothaan–Hall equations are an example.

Numerical Stability

Numerical stability can be a problem with this procedure and there are various ways of combating this instability. One of the most basic and generally applicable is called *F-mixing* or damping. With F-mixing, once a single electron wave function is calculated it is not used directly. Instead, some combination of that calculated wave function and the previous wave functions for that electron is used—the most common being a simple linear combination of the calculated and immediately preceding wave function. A clever dodge, employed by Hartree, for atomic calculations was to increase the nuclear charge, thus pulling all the electrons closer together. As the system stabilised, this was gradually reduced to the correct charge. In molecular calculations a similar approach is sometimes used by first calculating the wave function for a positive ion and then to use these orbitals as the starting point for the neutral molecule. Modern molecular Hartree–Fock computer programs use a variety of methods to ensure convergence of the Roothaan–Hall equations.

Weaknesses, Extensions, and Alternatives

Of the five simplifications outlined in the section "Hartree–Fock algorithm", the fifth is typically the most important. Neglect of electron correlation can lead to large deviations from experimental results. A number of approaches to this weakness, collectively called post-Hartree–Fock methods, have been devised to include electron correlation to the multi-electron wave function. One of these approaches, Møller–Plesset perturbation theory, treats correlation as a perturbation of the Fock operator. Others expand the true multi-electron wave function in terms of a linear combination of Slater determinants—such as multi-configurational self-consistent field, configuration interaction, quadratic configuration interaction, and complete active space SCF (CASSCF). Still others (such as variational quantum Monte Carlo) modify the Hartree–Fock wave function by multiplying it by a correlation function ("Jastrow" factor), a term which is explicitly a function of multiple electrons that cannot be decomposed into independent single-particle functions.

An alternative to Hartree–Fock calculations used in some cases is density functional theory, which treats both exchange and correlation energies, albeit approximately. Indeed, it is common to use calculations that are a hybrid of the two methods—the popular B3LYP scheme is one such hybrid functional method. Another option is to use modern valence bond methods.

Møller–Plesset Perturbation Theory

Møller–Plesset perturbation theory (MP) is one of several quantum chemistry post-Hartree–Fock ab initio methods in the field of computational chemistry. It improves on the Hartree–Fock method by adding electron correlation effects by means of Rayleigh–Schrödinger perturbation theory (RS-PT), usually to second (MP2), third

(MP3) or fourth (MP4) order. Its main idea was published as early as 1934 by Christian Møller and Milton S. Plesset.

Rayleigh–Schrödinger Perturbation Theory

The MP perturbation theory is a special case of RS perturbation theory. In RS theory one considers an unperturbed Hamiltonian operator $\hat{H}_0$, to which a small (often external) perturbation $\hat{V}$ is added:

$$\hat{H} = \hat{H}_0 + \lambda\hat{V}.$$

Here, λ is an arbitrary real parameter that controls the size of the perturbation. In MP theory the zeroth-order wave function is an exact eigenfunction of the Fock operator, which thus serves as the unperturbed operator. The perturbation is the correlation potential. In RS-PT the perturbed wave function and perturbed energy are expressed as a power series in λ:

$$\Psi = \lim_{m\to\infty} \sum_{i=0}^{m} \lambda^i \Psi^{(i)},$$

$$E = \lim_{m\to\infty} \sum_{i=0}^{m} \lambda^i E^{(i)}.$$

Substitution of these series into the time-independent Schrödinger equation gives a new equation as $m \to \infty$:

$$\left(\hat{H}_0 + \lambda V\right)\left(\sum_{i=0}^{m} \lambda^i \Psi^{(i)}\right) = \left(\sum_{i=0}^{m} \lambda^i E^{(i)}\right)\left(\sum_{i=0}^{m} \lambda^i \Psi^{(i)}\right).$$

Equating the factors of λ^k in this equation gives a kth-order perturbation equation, where $k = 0, 1, 2, ..., m$.

Møller–Plesset Perturbation

Original Formulation

The MP-energy corrections are obtained from Rayleigh–Schrödinger (RS) perturbation theory with the unperturbed Hamiltonian defined as the *shifted* Fock operator,

$$\hat{H}_0 \equiv \hat{F} + \langle \Phi_0 | (\hat{H} - \hat{F}) | \Phi_0 \rangle$$

and the perturbation defined as the *correlation potential*,

$$\hat{V} \equiv \hat{H} - \hat{H}_0 = \hat{H} - \left(\hat{F} + \langle \Phi_0 | (\hat{H} - \hat{F}) | \Phi_0 \rangle\right),$$

where the normalized Slater determinant Φ_0 is the lowest eigenstate of the Fock operator:

$$\hat{F}\Phi_0 \equiv \sum_{k=1}^{N} \hat{f}(k)\Phi_0 = 2\sum_{i=1}^{N/2}\varepsilon_i\Phi_0.$$

Here N is the number of electrons in the molecule under consideration (a factor of 2 in the energy arises from the fact that each orbital is occupied by a pair of electrons with opposite spin), $\hat{H}$ is the usual electronic Hamiltonian, $\hat{f}(k)$ is the one-electron Fock operator, and ε_i is the orbital energy belonging to the doubly occupied spatial orbital φ_i.

Since the Slater determinant Φ_0 is an eigenstate of $\hat{F}$, it follows readily that

$$\hat{F}\Phi_0 - \langle\Phi_0|\hat{F}|\Phi_0\rangle\Phi_0 = 0 \Rightarrow \hat{H}_0\Phi_0 = \langle\Phi_0|\hat{H}|\Phi_0\rangle\Phi_0,$$

i.e. the zeroth-order energy is the expectation value of $\hat{H}$ with respect to Φ_0, the Hartree-Fock energy. Similarly, it can been seen that *in this formulation* the MP1 energy

$$E_{MP1} \equiv \langle\Phi_0|\hat{V}|\Phi_0\rangle = 0.$$

Hence, the first meaningful correction appears at MP2 energy.

In order to obtain the MP2 formula for a closed-shell molecule, the second order RS-PT formula is written in a basis of doubly excited Slater determinants. (Singly excited Slater determinants do not contribute because of the Brillouin theorem). After application of the Slater–Condon rules for the simplification of N-electron matrix elements with Slater determinants in bra and ket and integrating out spin, it becomes

$$E_{MP2} = \frac{1}{4}\sum_{i,j,a,b}\frac{\langle\varphi_i\varphi_j|\hat{\tilde{v}}|\varphi_a\varphi_b\rangle\langle\varphi_a\varphi_b|\hat{\tilde{v}}|\varphi_i\varphi_j\rangle}{\varepsilon_i + \varepsilon_j - \varepsilon_a - \varepsilon_b}$$

where φ_i and φ_j are canonical occupied orbitals and φ_a and φ_b are virtual (or unoccupied) orbitals. The quantities ε_i, ε_j, ε_a, and ε_b are the corresponding orbital energies. Clearly, through second-order in the correlation potential, the total electronic energy is given by the Hartree–Fock energy plus second-order MP correction: $E \approx E_{HF} + E_{MP2}$. The solution of the zeroth-order MP equation (which by definition is the Hartree–Fock equation) gives the Hartree–Fock energy. The first non-vanishing perturbation correction beyond the Hartree–Fock treatment is the second-order energy.

Alternative Formulation

Equivalent expressions are obtained by a slightly different partitioning of the Hamiltonian, which results in a different division of energy terms over zeroth- and first-order contributions, while for second- and higher-order energy corrections the two partitionings give identical results. The formulation is commonly used by chemists, who are now large users of these methods. This difference is due to the fact, well known in Hartree–Fock theory, that

$$\langle \Phi_0 \,|\, (\hat{H} - \hat{F}) \,|\, \Phi_0 \rangle \neq 0 \quad \Leftrightarrow \quad E_{HF} \neq 2 \sum_{i=1}^{N/2} \varepsilon_i.$$

(The Hartree–Fock energy is *not* equal to the sum of occupied-orbital energies). In the alternative partitioning, one defines

$$\hat{H}_0 \equiv \hat{F}, \qquad \hat{V} \equiv \hat{H} - \hat{F}.$$

Clearly, in this partitioning,

$$E_{MP0} = 2 \sum_{i=1}^{N/2} \varepsilon_i, \qquad E_{MP1} = E_{HF} - 2 \sum_{i=1}^{N/2} \varepsilon_i.$$

Obviously, with this alternative formulation, the Møller–Plesset theorem does not hold in the literal sense that $E_{MP1} \neq 0$. The solution of the zeroth-order MP equation is the sum of orbital energies. The zeroth plus first-order correction yields the Hartree–Fock energy. As with the original formulation, the first non-vanishing perturbation correction beyond the Hartree–Fock treatment is the second-order energy. To reiterate, the second- and higher-order corrections are the same in both formulations.

Use of Møller–Plesset Perturbation Methods

Second (MP2), third (MP3), and fourth (MP4) order Møller–Plesset calculations are standard levels used in calculating small systems and are implemented in many computational chemistry codes. Higher level MP calculations, generally only MP5, are possible in some codes. However, they are rarely used because of their cost.

Systematic studies of MP perturbation theory have shown that it is not necessarily a convergent theory at high orders. Convergence can be slow, rapid, oscillatory, regular, highly erratic or simply non-existent, depending on the precise chemical system or basis set. The density matrix for the first-order and higher MP2 wavefunction is of the type known as *response density*, which differs from the more usual *expectation value density*. The eigenvalues of the response density matrix (which are the occupation numbers of the MP2 natural orbitals) can therefore be greater than 2 or negative. Unphysical numbers are a sign of a divergent perturbation expansion.

Additionally, various important molecular properties calculated at MP3 and MP4 level are no better than their MP2 counterparts, even for small molecules.

For open shell molecules, MPn-theory can directly be applied only to unrestricted Hartree–Fock reference functions (since ROHF states are not in general eigenvectors of the Fock operator). However, the resulting energies often suffer from severe spin contamination, leading to large errors. A possible better alternative is to use one of the MP2-like methods based on restricted open-shell Hartree–Fock (ROHF). Unfortunately, there are many ROHF based MP2-like methods because of arbitrariness in the ROHF wavefunction(for example HCPT, ROMP, RMP (also called ROHF-MBPT2), OPT1 and OPT2, ZAPT, IOPT, etc.). Some of the ROHF based MP2-like theories suffer from spin-contamination in their perturbed density and energies beyond second-order.

These methods, Hartree–Fock, unrestricted Hartree–Fock and restricted Hartree–Fock use a single determinant wave function. Multi-configurational self-consistent field (MCSCF) methods use several determinants and can be used for the unperturbed operator, although not uniquely, so many methods, such as complete active space perturbation theory (CASPT2), and Multi-Configuration Quasi-Degenerate Perturbation Theory (MCQDPT), have been developed. Unfortunately, MCSCF based methods are not without perturbation series divergences.

Car–Parrinello Molecular Dynamics

Car–Parrinello molecular dynamics or CPMD refers to either a method used in molecular dynamics (also known as the Car–Parrinello method) or the computational chemistry software package used to implement this method.

The CPMD method is related to the more common Born–Oppenheimer molecular dynamics (BOMD) method in that the quantum mechanical effect of the electrons is included in the calculation of energy and forces for the classical motion of the nuclei. However, whereas BOMD treats the electronic structure problem within the time-*independent* Schrödinger equation, CPMD explicitly includes the electrons as active degrees of freedom, via (fictitious) dynamical variables.

The software is a parallelized plane wave/pseudopotential implementation of density functional theory, particularly designed for *ab initio* molecular dynamics.

Car–Parrinello Method

The Car–Parrinello method is a type of molecular dynamics, usually employing periodic boundary conditions, planewave basis sets, and density functional theory, proposed by Roberto Car and Michele Parrinello in 1985, who were subsequently awarded the Dirac Medal by ICTP in 2009.

In contrast to Born–Oppenheimer molecular dynamics wherein the nuclear (ions) degree of freedom are propagated using ionic forces which are calculated at each iteration by approximately solving the electronic problem with conventional matrix diagonalization methods, the Car–Parrinello method explicitly introduces the electronic degrees of freedom as (fictitious) dynamical variables, writing an extended Lagrangian for the system which leads to a system of coupled equations of motion for both ions and electrons. In this way an explicit electronic minimization at each time step, as done in Born-Oppenheimer MD, is not needed: after an initial standard electronic minimization, the fictitious dynamics of the electrons keeps them on the electronic ground state corresponding to each new ionic configuration visited along the dynamics, thus yielding accurate ionic forces. In order to maintain this adiabaticity condition, it is necessary that the fictitious mass of the electrons is chosen small enough to avoid a significant energy transfer from the ionic to the electronic degrees of freedom. This small fictitious mass in turn requires that the equations of motion are integrated using a smaller time step than the one (1–10 fs) commonly used in Born–Oppenheimer molecular dynamics.

General Approach

In CPMD the core electrons are usually described by a pseudopotential and the wavefunction of the valence electrons are approximated by a plane wave basis set.

The ground state electronic density (for fixed nuclei) is calculated self-consistently, usually using the density functional theory method. Then, using that density, forces on the nuclei can be computed, to update the trajectories (using, e.g. the Verlet integration algorithm). In addition, however, the coefficients used to obtain the electronic orbital functions can be treated as a set of extra spatial dimensions, and trajectories for the orbitals can be calculated in this context.

Fictitious Dynamics

CPMD is an approximation of the Born–Oppenheimer MD (BOMD) method. In BOMD, the electrons' wave function must be minimized via matrix diagonalization at every step in the trajectory. CPMD uses fictitious dynamics to keep the electrons close to the ground state, preventing the need for a costly self-consistent iterative minimization at each time step. The fictitious dynamics relies on the use of a fictitious electron mass (usually in the range of $400 - 800$ a.u.) to ensure that there is very little energy transfer from nuclei to electrons, i.e. to ensure adiabaticity. Any increase in the fictitious electron mass resulting in energy transfer would cause the system to leave the ground-state BOMD surface.

Lagrangian

$$\mathcal{L} = \frac{1}{2}\left(\sum_{I}^{\text{nuclei}} M_I \, \dot{\mathbf{R}}_I^2 + \mu \sum_{i}^{\text{orbitals}} \int d\mathbf{r} \, |\dot{\psi}_i(\mathbf{r},t)|^2 \right) - E\big[\{\psi_i\},\{\mathbf{R}_I\}\big],$$

where $E[\{\psi_i\},\{R_I\}]$ is the Kohn–Sham energy density functional, which outputs energy values when given Kohn–Sham orbitals and nuclear positions.

Orthogonality Constraint

$$\int d\mathbf{r}\, \psi_i^*(\mathbf{r},t)\psi_j(\mathbf{r},t) = \delta_{ij},$$

where δ_{ij} is the Kronecker delta.

Equations of Motion

The equations of motion are obtained by finding the stationary point of the Lagrangian under variations of ψ_i and R_I, with the orthogonality constraint.

$$M_I\,\ddot{\mathbf{R}}_I = -\nabla_I E\left[\{\psi_i\},\{\mathbf{R}_J\}\right]$$

$$\mu\ddot{\psi}_i(\mathbf{r},t) = -\frac{\delta E}{\delta\psi_i^*(\mathbf{r},t)} + \sum_j \Lambda_{ij}\psi_j(\mathbf{r},t),$$

where Λ_{ij} is a Lagrangian multiplier matrix to comply with the orthonormality constraint.

Born–Oppenheimer Limit

In the formal limit where $\mu \to 0$, the equations of motion approach Born–Oppenheimer molecular dynamics.

Application

1. Studying the behavior of water near a hydrophobic graphene sheet.

2. Investigating the structure of liquid water at ambient temperature.

3. Solving the heat transfer problems (heat conduction and thermal radiation) between Si/Ge superlattices.

4. Probing the proton transfer along 1D water chains inside carbon nanotubes.

5. Evaluating the critical point of aluminum.

Matched Molecular Pair Analysis

Matched molecular pair analysis (MMPA) is a method in cheminformatics whereby the pedigree of structural changes, within drug like small molecules, is established based on real measured data. The term was first coined by Kenny and Sadowski in the book *Chemoinformatics in Drug Discovery*. The basic notion of MMPA based analysis is analyzing chemical data sets dealing with pairs of compounds. Such pairs of compounds are known as matched molecular pairs (MMP).

Introduction

MMP can be defined as a pair of molecules that differ in only a minor single point change. Matched molecular pairs (MMPs) are widely used in medicinal chemistry to study changes in compound properties which includes biological activity, toxicity, environmental hazards and much more, which are associated with well-defined structural modifications. Single point changes in the molecule pairs are termed a chemical transformation or Molecular transformation. Each molecular pair is associated with a particular transformation. An example of transformation is the replacement of one functional group by another. More specifically, molecular transformation can be defined as the replacement of a molecular fragment having one, two or three attachment points with another fragment. Useful Molecular transformation in a specified context is termed as "Significant" transformations. For example, a transformation may systematically decrease or increase a desired property of chemical compounds. Transformations that affect a particular property/activity in a statistically significant sense are called as significant transformations. The transformation is considered significant, if it increases the property value "more often" than it decreases it or vice versa. Thus, the distribution of increasing and decreasing pairs should be significantly different from the binomial ("no effect") distribution with a particular p-value (usually 0.05).

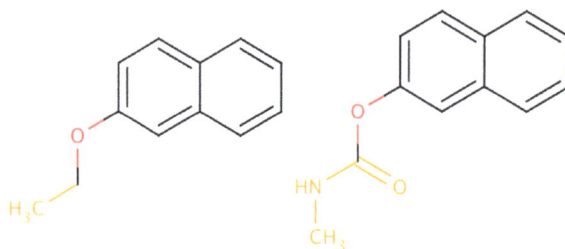

Examplary MMPs (differences highlighted in orange)

Significance of MMP Based Analysis

MMP based analysis is an attractive method for computational analysis because they can be algorithmically generated and they make it possible to associate defined structural modifications at the level of compound pairs with chemical property changes, including biological activity.

Interpretable QSAR Models

MMPA is quite useful in the field of quantitative structure–activity relationship (QSAR) modelling studies. One of the issues of QSAR models is they are difficult to interpret in a chemically meaningful manner. While it can be pretty easy to interpret simple linear regression models, the most powerful algorithms like neural networks, support vector machine are similar to "black boxes", which provide predictions that can't be easily interpreted. This problem undermines the applicability of QSAR model in helping the medicinal chemist to make the decision. If the compound is predicted to be active against some microorganism, what are the driving factors of its activity? Or if it is predicted to be inactive, how its activity can be modulated? The black box nature of the QSAR model prevents it from addressing these crucial issues. The use of predicted MMPs allows to interpret models and identify which MMPs were learned by the model. The MMPs, which were not reproduced by the model, could correspond to experimental errors or deficiency of the model (inappropriate descriptors, too few data, etc.).

Analysis of MMPs (matched molecular pair) can be very useful for understanding the mechanism of action. A medicinal chemist might be interested particularly in "activity cliff". Activity cliff is a minor structural modification, which changes the target activity significantly.

Activity Cliff

Activity cliffs are minor structural modifications, with significant effect on molecular property. Activity cliffs usually have high SAR information content. Because small chemical changes in the set of similar compounds lead to large changes in activity. The assessment of activity cliffs requires careful consideration of similarity and potency difference criteria.

Types of MMP Based Analysis

Matched molecular pair (MMPA) analyses can be classified into two types: supervised and unsupervised MMPA.

Supervised MMPA

In supervised MMPA, the chemical transformations are predefined, then the corresponding matched pair compounds are found within the data set and the change in end point computed for each transformation.

Unsupervised MMPA

Also known as automated MMPAs. A machine learning algorithm is used to finds all possible matched pairs in a data set according to a set of predefined rules. This results in much larger numbers of matched pairs and unique transformations, which are typically filtered during the process to identify those transformations that correspond to statistically significant changes in the targeted property with a reasonable number of matched pairs.

Matched Molecular Series

Here instead of looking at the pair of molecules which differ only at one point, a series of more than 2 molecules different at a single point is considered. The concept of matching molecular series was introduced by Wawer and Bajorath. It is argued that longer matched series is more likely to exhibit preferred molecular transformation while, matched pairs exhibit only a small preference.

Limitations

The application of the MMPA across large chemical databases for the optimization of ligand potency is problematic because same structural transformation may increase or decrease or doesn't affect the potency of different compounds in the dataset. Selection of practical significant transformation from a dataset of molecules is a challenging issue in the MMPA. Moreover, the effect of a particular molecular transformation can significantly depend on the Chemical context of transformations.

Beside these, MMPA might pose some limitations in terms of computational resources, especially when dealing with databases of compounds with a large number of breakable bonds. Further, more atoms in the variable part of the molecule also leads to combinatorial explosion problems. To deal with this, the number of breakable bonds and number of atoms in the variable part can be used to pre-filter the database.

References

- Sínanoğlu, Oktay (1962). "Many-Electron Theory of Atoms and Molecules. I. Shells, Electron Pairs vs Many-Electron Correlations". The Journal of Chemical Physics. 36 (3): 706. Bibcode:-1962JChPh..36..706S. doi:10.1063/1.1732596

- Jensen, Frank (2007). Introduction to Computational Chemistry. Chichester, England: John Wiley and Sons. pp. 80–81. ISBN 0-470-01187-4

- Chen, T. C. (1955). "Expansion of Electronic Wave Functions of Molecules in Terms of 'United-Atom' Wave Functions". J. Chem. Phys. 23 (11): 2200–2201. doi:10.1063/1.1740713

- Curtiss, Larry A.; Paul C. Redfern; Krishan Raghavachari (2007). "Gaussian-4 theory". The Journal of Chemical Physics. 126 (8): 084108. Bibcode:2007JChPh.126h4108C. PMID 17343441. doi:10.1063/1.2436888

- Cramer, Christopher J. (2002). Essentials of Computational Chemistry. Chichester: John Wiley & Sons, Ltd. pp. 191–232. ISBN 0-471-48552-7

- "Ab initio study of phase stability in doped TiO_2". Computational Mechanics. 50 (2): 185–194. 2012. doi:10.1007/s00466-012-0728-4

- Michelini, M. C.; Pis Diez, R.; Jubert, A. H. (25 June 1998). "A Density Functional Study of Small Nickel Clusters". International Journal of Quantum Chemistry. 70 (4–5): 694. doi:10.1002/(SICI)1097-461X(1998)70:4/5<693::AID-QUA15>3.0.CO;2-3. Retrieved 21 October 2016

- Kieron Burke; Lucas O. Wagner (2013). "DFT in a nutshell". International Journal of Quantum Chemistry. 113 (2): 96. doi:10.1002/qua.24259

- Hinchliffe, Alan (2000). Modelling Molecular Structures (2nd ed.). Baffins Lane, Chichester, West Sussex PO19 1UD, England: John Wiley & Sons Ltd. p. 186. ISBN 0-471-48993-X

- James J. P. Stewart (1989). "Optimization of parameters for semiempirical methods I. Method". The Journal of Computational Chemistry. Wiley InterScience. 10 (2): 209–220. doi:10.1002/jcc.540100208

- Davidson, Ernest R.; Jarzecki, A. A. (1999). K. Hirao, ed. Recent Advances in Multireference Methods. World Scientific. pp. 31–63. ISBN 981-02-3777-4

- Hagen, G.; Papenbrock, T.; Hjorth-Jensen, M.; Dean, D. J. (2014). "Coupled-cluster computations of atomic nuclei". Reports on Progress in Physics. 77 (9): 096302. doi:10.1088/0034-4885/77/9/096302

- Glaesemann, Kurt R.; Schmidt, Michael W. (2010). "On the Ordering of Orbital Energies in High-Spin ROHF+". The Journal of Physical Chemistry A. 114 (33): 8772–8777. PMID 20443582. doi:10.1021/jp101758y

- M. Brack (1983), "Virial theorems for relativistic spin-½ and spin-0 particles", Phys. Rev. D, 27: 1950, doi:10.1103/physrevd.27.1950

- Tong, Lianheng. "Methfessel-Paxton Approximation to Step Function". Metal CONQUEST. Retrieved 21 October 2016

Understanding Molecular Dynamics

Molecular dynamics helps in developing a better understanding of the movement of atoms and molecules. Computer simulation is used in this process. Tools and techniques are an important component of any field of study. The following chapter elucidates the various tools and techniques that are related to molecular dynamics.

Molecular Dynamics

Molecular dynamics (MD) is a computer simulation method for studying the physical movements of atoms and molecules, and is thus a type of N-body simulation. The atoms and molecules are allowed to interact for a fixed period of time, giving a view of the dynamic evolution of the system. In the most common version, the trajectories of atoms and molecules are determined by numerically solving Newton's equations of motion for a system of interacting particles, where forces between the particles and their potential energies are calculated using interatomic potentials or molecular mechanics force fields. The method was originally developed within the field of theoretical physics in the late 1950s but is applied today mostly in chemical physics, materials science and the modelling of biomolecules.

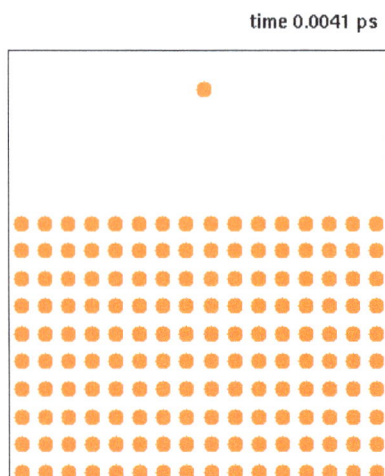

time 0.0041 ps

Example of a molecular dynamics simulation in a simple system: deposition of one copper (Cu) atom on a Cu Miller index (001) surface. Each circle illustrates the position of one atom. The atomic interactions used in current simulations are more complex than those of 2-dimensional hard spheres.

Simplified schematic of the molecular dynamics algorithm

Give atoms initial $r^{(i=0)}$ and $v^{(i=0)}$, set $a = 0.0$, $t = 0.0$, $i = 0$, choose short Δt

Predictor stage: predict next atom positions: Move atoms: $r^p = r^{(i)} + v^{(i)} \Delta t + \frac{1}{2} a \Delta t^2$ + more accurate terms Update velocities: $v^p = v^{(i)} + a \Delta t$ + more accurate terms

Get forces $F = - \nabla V(r^p)$ or $F = F(\Psi(r^p))$ and $a = F/m$

Corrector stage: adjust atom positions based on new a: Move atoms: $r^{(i+1)} = r^p$ + some function of $(a, \Delta t)$ Update velocities: $v^{(i+1)} = v^p$ + some function of $(a, \Delta t)$

Apply boundary conditions, temperature and pressure control as needed

Calculate and output physical quantities of interest

Move time and iteration step forward: $t = t + \Delta t$, $i = i + 1$

Repeat as long as you need

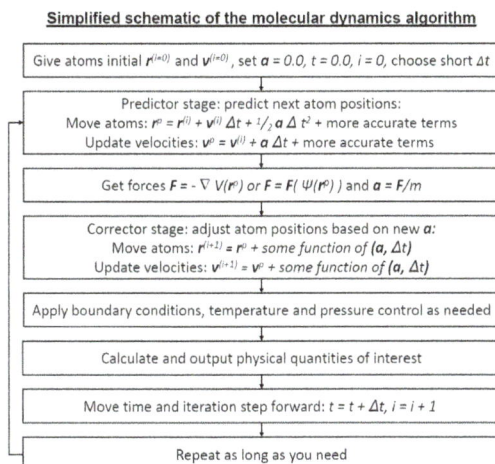

A simplified description of the standard molecular dynamics simulation algorithm, when a predictor-corrector-type integrator is used. The forces may come either from classical interatomic potentials (indicated as $F = -\nabla V(\vec{r})$) or quantum mechanical (indicated schematically as $F = F(\Psi(\vec{r}))$) methods. Large differences exist between different integrators; some do not exactly have the same highest-order terms as indicated in the schematic, many use also higher-order time derivatives, and some use both the current and prior time step in variable-time step schemes.

Because molecular systems typically consist of a vast number of particles, it is impossible to determine the properties of such complex systems analytically; MD simulation circumvents this problem by using numerical methods. However, long MD simulations are mathematically ill-conditioned, generating cumulative errors in numerical integration that can be minimized with proper selection of algorithms and parameters, but not eliminated entirely.

For systems which obey the ergodic hypothesis, the evolution of one molecular dynamics simulation may be used to determine macroscopic thermodynamic properties of the system: the time averages of an ergodic system correspond to microcanonical ensemble averages. MD has also been termed "statistical mechanics by numbers" and "Laplace's vision of Newtonian mechanics" of predicting the future by animating nature's forces and allowing insight into molecular motion on an atomic scale.

History

Following the earlier successes of Monte Carlo simulations, the method was developed by Fermi, Pasta, and Ulam (and Tsingou) in the mid 50s, Alder and Wainwright in late 50s and Rahman (independently) in the 60s. In 1957, Alder and Wainwright used an IBM 704 computer to simulate perfectly elastic collisions between hard spheres. In 1960, Gibson et al., simulated radiation damage of solid copper by using a Born-Mayer type of repulsive interaction along with a cohesive surface force. In 1964, Rahman published landmark simulations of liquid argon that used a Lennard-Jones potential. Calculations of system properties, such as the coefficient of self-diffusion, compared well with experimental data.

Even before it became possible to simulate molecular dynamics with computers, some undertook the hard work of trying it with physical models such as macroscopic spheres. The idea was to arrange them to replicate the properties of a liquid. J.D. Bernal said, in 1962: "*... I took a number of rubber balls and stuck them together with rods of a selection of different lengths ranging from 2.75 to 4 inches. I tried to do this in the first place as casually as possible, working in my own office, being interrupted every five minutes or so and not remembering what I had done before the interruption.*"

Areas of Application and Limits

Beginning in theoretical physics, the method of MD gained popularity in materials science and since the 1970s also in biochemistry and biophysics. MD is frequently used to refine three-dimensional structures of proteins and other macromolecules based on experimental constraints from X-ray crystallography or NMR spectroscopy. In physics, MD is used to examine the dynamics of atomic-level phenomena that cannot be observed directly, such as thin film growth and ion-subplantation. It is also used to examine the physical properties of nanotechnological devices that have not been or cannot yet be created. In biophysics and structural biology, the method is frequently applied to study the motions of biological macromolecules such as proteins and nucleic acids, which can be useful for interpreting the results of certain biophysical experiments and for modeling interactions with other molecules, as in ligand docking. In principle MD can be used for ab initio prediction of protein structure by simulating folding of the polypeptide chain from random coil.

The results of MD simulations can be tested through comparison to experiments that measure molecular dynamics, of which a popular method is nuclear magnetic resonance spectroscopy. MD-derived structure predictions can be tested through community-wide experiments in Critical Assessment of protein Structure Prediction (CASP), although the method has historically had limited success in this area. Michael Levitt, who shared the Nobel Prize awarded in part for the application of MD to proteins, wrote in 1999 that CASP participants usually did not use the method due to "*... a central embarrassment of molecular mechanics, namely that energy minimization or molecular dynamics generally leads to a model that is less like the experimental structure.*" Improvements in computational resources permitting more and longer MD trajectories to be calculated, combined with modern improvements in the quality of force field parameters, have yielded some improvements in performance of both structure prediction and homology model refinement, without reaching the point of practical utility in these areas; most such work identifies force field parameters as a key area for further development.

Limits of the method are related to the parameter sets used, and to the underlying molecular mechanics force fields. One run of an MD simulation optimizes the potential energy, rather than the free energy of the protein, meaning that all entropic contributions to thermodynamic stability of protein structure are neglected. The neglected contributions include the conformational entropy of the polypeptide chain (which is the main factor that destabilizes protein structure) and hydrophobic effects that are the main driving forces of protein folding. Another important factor are intramolecular hydrogen bonds, which are not explicitly included in modern force fields, but described as Coulomb interactions of atomic point charges. This is a crude approximation because hydrogen bonds have a partially quantum mechanical and chemical nature. Furthermore, electrostatic interactions are usually calculated using the dielectric constant of vacuum, although the surrounding aqueous solution has a much higher dielectric constant. Using the macroscopic dielectric constant at short interatomic distances is questionable. Finally, van der Waals interactions in MD are usually described by Lennard-Jones potentials based on the Fritz London theory that is only applicable in vacuum. However, all types of van der Waals forces are ultimately of electrostatic origin and therefore depend on dielectric properties of the environment. The direct measurement of attraction forces between different materials (as Hamaker constant) shows that "the interaction between hydrocarbons across water is about 10% of that across vacuum". The environment-dependence of van der Waals forces is neglected in standard simulations, but can be included by developing polarizable force fields.

Design Constraints

Design of a molecular dynamics simulation should account for the available computational power. Simulation size (n=number of particles), timestep, and total time duration must be selected so that the calculation can finish within a reasonable time period. However, the simulations should be long enough to be relevant to the time scales of the natural processes being studied. To make statistically valid conclusions from the simulations, the time span simulated should match the kinetics of the natural process. Otherwise, it is analogous to making conclusions about how a human walks when only looking at less than one footstep. Most scientific publications about the dynamics of proteins and DNA use data from simulations spanning nanoseconds (10^{-9} s) to microseconds (10^{-6} s). To obtain these simulations, several CPU-days to CPU-years are needed. Parallel algorithms allow the load to be distributed among CPUs; an example is the spatial or force decomposition algorithm.

During a classical MD simulation, the most CPU intensive task is the evaluation of the potential as a function of the particles' internal coordinates. Within that energy evaluation, the most expensive one is the non-bonded or non-covalent part. In Big O notation, common molecular dynamics simulations scale by $O(n^2)$ if all pair-wise electrostatic and van der Waals interactions must be accounted for explicitly. This computational cost can be reduced by employing electrostatics methods such as particle mesh Ewald summation ($O(n \log(n))$), particle–particle-particle–mesh (P3M), or good spherical cutoff methods ($O(n)$).

Another factor that impacts total CPU time needed by a simulation is the size of the integration timestep. This is the time length between evaluations of the potential. The timestep must be chosen small enough to avoid discretization errors (i.e., smaller than the fastest vibrational frequency in the system). Typical timesteps for classical MD are in the order of 1 femtosecond (10^{-15} s). This value may be extended by using algorithms such as the SHAKE constraint algorithm, which fix the vibrations of the fastest atoms (e.g., hydrogens) into place. Multiple time scale methods have also been developed, which allow extended times between updates of slower long-range forces.

For simulating molecules in a solvent, a choice should be made between explicit solvent and implicit solvent. Explicit solvent particles (such as the TIP3P, SPC/E and SPC-f water models) must be calculated expensively by the force field, while implicit solvents use a mean-field approach. Using an explicit solvent is computationally expensive, requiring inclusion of roughly ten times more particles in the simulation. But the granularity and viscosity of explicit solvent is essential to reproduce certain properties of the solute molecules. This is especially important to reproduce chemical kinetics.

In all kinds of molecular dynamics simulations, the simulation box size must be large enough to avoid boundary condition artifacts. Boundary conditions are often treated by choosing fixed values at the edges (which may cause artifacts), or by employing periodic boundary conditions in which one side of the simulation loops back to the opposite side, mimicking a bulk phase.

Microcanonical Ensemble (NVE)

In the *microcanonical* (NVE) ensemble, the system is isolated from changes in moles (N), volume (V), and energy (E). It corresponds to an adiabatic process with no heat exchange. A microcanonical molecular dynamics trajectory may be seen as an exchange of potential and kinetic energy, with total energy being conserved. For a system of N particles with coordinates X and velocities V, the following pair of first order differential equations may be written in Newton's notation as

$$F(X) = -\nabla U(X) = M\dot{V}(t)$$

$$V(t) = \dot{X}(t).$$

The potential energy function $U(X)$ of the system is a function of the particle coordinates X. It is referred to simply as the *potential* in physics, or the *force field* in chemistry. The first equation comes from Newton's laws of motion; the force F acting on each particle in the system can be calculated as the negative gradient of $U(X)$.

For every time step, each particle's position X and velocity V may be integrated with a symplectic integrator method such as Verlet integration. The time evolution of X and V is called a trajectory. Given the initial positions (e.g., from theoretical knowledge) and velocities (e.g., randomized Gaussian), we can calculate all future (or past) positions and velocities.

One frequent source of confusion is the meaning of temperature in MD. Commonly we have experience with macroscopic temperatures, which involve a huge number of particles. But temperature is a statistical quantity. If there is a large enough number of atoms, statistical temperature can be estimated from the *instantaneous temperature*, which is found by equating the kinetic energy of the system to $nk_BT/2$ where n is the number of degrees of freedom of the system.

A temperature-related phenomenon arises due to the small number of atoms that are used in MD simulations. For example, consider simulating the growth of a copper film starting with a substrate containing 500 atoms and a deposition energy of 100 eV. In the real world, the 100 eV from the deposited atom would rapidly be transported through and shared among a large number of atoms (10^{10} or more) with no big change in temperature. When there are only 500 atoms, however, the substrate is almost immediately vaporized by the deposition. Something similar happens in biophysical simulations. The temperature of the system in NVE is naturally raised when macromolecules such as proteins undergo exothermic conformational changes and binding.

Canonical Ensemble (NVT)

In the canonical ensemble, amount of substance (N), volume (V) and temperature (T) are conserved. It is also sometimes called constant temperature molecular dynamics (CTMD). In NVT, the energy of endothermic and exothermic processes is exchanged with a thermostat.

A variety of thermostat algorithms are available to add and remove energy from the boundaries of a MD simulation in a more or less realistic way, approximating the canonical ensemble. Popular methods to control temperature include velocity rescaling, the Nosé-Hoover thermostat, Nosé-Hoover chains, the Berendsen thermostat, the Andersen thermostat and Langevin dynamics. The Berendsen thermostat might introduce the flying ice cube effect, which leads to unphysical translations and rotations of the simulated system.

It is not trivial to obtain a canonical ensemble distribution of conformations and velocities using these algorithms. How this depends on system size, thermostat choice, thermostat parameters, time step and integrator is the subject of many articles in the field.

Isothermal–isobaric (NPT) Ensemble

In the isothermal–isobaric ensemble, amount of substance (N), pressure (P) and temperature (T) are conserved. In addition to a thermostat, a barostat is needed. It corresponds most closely to laboratory conditions with a flask open to ambient temperature and pressure.

In the simulation of biological membranes, isotropic pressure control is not appropriate. For lipid bilayers, pressure control occurs under constant membrane area (NPAT) or constant surface tension "gamma" (NPγT).

Generalized Ensembles

The replica exchange method is a generalized ensemble. It was originally created to deal with the slow dynamics of disordered spin systems. It is also called parallel tempering. The replica exchange MD (REMD) formulation tries to overcome the multiple-minima problem by exchanging the temperature of non-interacting replicas of the system running at several temperatures.

Potentials in MD Simulations

A molecular dynamics simulation requires the definition of a potential function, or a description of the terms by which the particles in the simulation will interact. In chemistry and biology this is usually referred to as a force field and in materials physics as an interatomic potential. Potentials may be defined at many levels of physical accuracy; those most commonly used in chemistry are based on molecular mechanics and embody a classical mechanics treatment of particle-particle interactions that can reproduce structural and conformational changes but usually cannot reproduce chemical reactions.

The reduction from a fully quantum description to a classical potential entails two main approximations. The first one is the Born–Oppenheimer approximation, which states that the dynamics of electrons are so fast that they can be considered to react instantaneously to the motion of their nuclei. As a consequence, they may be treated separately. The second one treats the nuclei, which are much heavier than electrons, as point particles that follow classical Newtonian dynamics. In classical molecular dynamics, the effect of the electrons is approximated as one potential energy surface, usually representing the ground state.

When finer levels of detail are needed, potentials based on quantum mechanics are used; some methods attempt to create hybrid classical/quantum potentials where the bulk of the system is treated classically but a small region is treated as a quantum system, usually undergoing a chemical transformation.

Empirical Potentials

Empirical potentials used in chemistry are frequently called force fields, while those used in materials physics are called interatomic potentials.

Most force fields in chemistry are empirical and consist of a summation of bonded forces associated with chemical bonds, bond angles, and bond dihedrals, and non-bonded forces associated with van der Waals forces and electrostatic charge. Empirical potentials represent quantum-mechanical effects in a limited way through ad-hoc functional approximations. These potentials contain free parameters such as atomic charge, van der Waals parameters reflecting estimates of atomic radius, and equilibrium bond length, angle, and dihedral; these are obtained by fitting against detailed electronic calculations (quantum chemical simulations) or experimental physical properties such as elastic constants, lattice parameters and spectroscopic measurements.

Because of the non-local nature of non-bonded interactions, they involve at least weak interactions between all particles in the system. Its calculation is normally the bottleneck in the speed of MD simulations. To lower the computational cost, force fields employ numerical approximations such as shifted cutoff radii, reaction field algorithms, particle mesh Ewald summation, or the newer particle–particle-particle–mesh (P3M).

Chemistry force fields commonly employ preset bonding arrangements (an exception being *ab initio* dynamics), and thus are unable to model the process of chemical bond breaking and reactions explicitly. On the other hand, many of the potentials used in physics, such as those based on the bond order formalism can describe several different coordinations of a system and bond breaking. Examples of such potentials include the Brenner potential for hydrocarbons and its further developments for the C-Si-H and C-O-H systems. The ReaxFF potential can be considered a fully reactive hybrid between bond order potentials and chemistry force fields.

Pair Potentials Versus Many-body Potentials

The potential functions representing the non-bonded energy are formulated as a sum over interactions between the particles of the system. The simplest choice, employed in many popular force fields, is the "pair potential", in which the total potential energy can be calculated from the sum of energy contributions between pairs of atoms. An example of such a pair potential is the non-bonded Lennard–Jones potential (also termed the 6–12 potential), used for calculating van der Waals forces.

$$U(r) = 4\varepsilon \left[\left(\frac{\sigma}{r} \right)^{12} - \left(\frac{\sigma}{r} \right)^{6} \right]$$

Another example is the Born (ionic) model of the ionic lattice. The first term in the next equation is Coulomb's law for a pair of ions, the second term is the short-range repulsion explained by Pauli's exclusion principle and the final term is the dispersion interaction term. Usually, a simulation only includes the dipolar term, although sometimes the quadrupolar term is also included.(Usually termed Buckingham potential model)

$$U_{ij}(r_{ij}) = \frac{z_i z_j}{4\pi\epsilon_0} \frac{1}{r_{ij}} + A_l \exp\frac{-r_{ij}}{p_l} + C_l r_{ij}^{-n_l} + \cdots$$

In many-body potentials, the potential energy includes the effects of three or more particles interacting with each other. In simulations with pairwise potentials, global interactions in the system also exist, but they occur only through pairwise terms. In many-body potentials, the potential energy cannot be found by a sum over pairs of atoms, as these interactions are calculated explicitly as a combination of higher-order terms. In the statistical view, the dependency between the variables cannot in general be expressed using only pairwise products of the degrees of freedom. For example, the Tersoff potential, which was originally used to simulate carbon, silicon, and germanium, and has since been used for a wide range of other materials, involves a sum over groups of three atoms, with the angles between the atoms being an important factor in the potential. Other examples are the embedded-atom method (EAM), the EDIP, and the Tight-Binding Second Moment Approximation (TBSMA) potentials, where the electron density of states in the region of an atom is calculated from a sum of contributions from surrounding atoms, and the potential energy contribution is then a function of this sum.

Semi-empirical Potentials

Semi-empirical potentials make use of the matrix representation from quantum mechanics. However, the values of the matrix elements are found through empirical formulae that estimate the degree of overlap of specific atomic orbitals. The matrix is then diagonalized to determine the occupancy of the different atomic orbitals, and empirical formulae are used once again to determine the energy contributions of the orbitals.

There are a wide variety of semi-empirical potentials, termed tight-binding potentials, which vary according to the atoms being modeled.

Polarizable Potentials

Most classical force fields implicitly include the effect of polarizability, e.g., by scaling up the partial charges obtained from quantum chemical calculations. These partial charges are stationary with respect to the mass of the atom. But molecular dynamics simulations can explicitly model polarizability with the introduction of induced dipoles through different methods, such as Drude particles or fluctuating charges. This allows for a dynamic redistribution of charge between atoms which responds to the local chemical environment.

For many years, polarizable MD simulations have been touted as the next generation. For homogenous liquids such as water, increased accuracy has been achieved through the inclusion of polarizability. Some promising results have also been achieved for proteins. However, it is still uncertain how to best approximate polarizability in a simulation.

Potentials in ab Initio Methods

In classical molecular dynamics, one potential energy surface (usually the ground state) is represented in the force field. This is a consequence of the Born–Oppenheimer approximation. In excited states, chemical reactions or when a more accurate representation is needed, electronic behavior can be obtained from first principles by using a quantum mechanical method, such as density functional theory. This is named Ab Initio Molecular Dynamics (AIMD). Due to the cost of treating the electronic degrees of freedom, the computational cost of these simulations is far higher than classical molecular dynamics. This implies that AIMD is limited to smaller systems and shorter times.

Ab initio quantum mechanical and chemical methods may be used to calculate the potential energy of a system on the fly, as needed for conformations in a trajectory. This calculation is usually made in the close neighborhood of the reaction coordinate. Although various approximations may be used, these are based on theoretical considerations, not on empirical fitting. *Ab initio* calculations produce a vast amount of information that is not available from empirical methods, such as density of electronic states or other electronic properties. A significant advantage of using *ab initio* methods is the ability to study reactions that involve breaking or formation of covalent bonds, which correspond to multiple electronic states.

Hybrid QM/MM

QM (quantum-mechanical) methods are very powerful. However, they are computationally expensive, while the MM (classical or molecular mechanics) methods are fast but suffer from several limits (require extensive parameterization; energy estimates obtained are not very accurate; cannot be used to simulate reactions where covalent bonds are broken/formed; and are limited in their abilities for providing accurate details regarding the chemical environment). A new class of method has emerged that combines the good points of QM (accuracy) and MM (speed) calculations. These methods are termed mixed or hybrid quantum-mechanical and molecular mechanics methods (hybrid QM/MM).

The most important advantage of hybrid QM/MM method is the speed. The cost of doing classical molecular dynamics (MM) in the most straightforward case scales $O(n^2)$, where n is the number of atoms in the system. This is mainly due to electrostatic interactions term (every particle interacts with every other particle). However, use of cutoff radius, periodic pair-list updates and more recently the variations of the particle-mesh Ewald's (PME) method has reduced this to between $O(n)$ to $O(n^2)$. In other words, if a system with twice as many atoms is simulated then it would take between two and four times as much computing power. On the other hand, the simplest *ab initio* calculations typically scale $O(n^3)$

or worse (restricted Hartree–Fock calculations have been suggested to scale ~$O(n^{2.7})$). To overcome the limit, a small part of the system is treated quantum-mechanically (typically active-site of an enzyme) and the remaining system is treated classically.

In more sophisticated implementations, QM/MM methods exist to treat both light nuclei susceptible to quantum effects (such as hydrogens) and electronic states. This allows generating hydrogen wave-functions (similar to electronic wave-functions). This methodology has been useful in investigating phenomena such as hydrogen tunneling. One example where QM/MM methods have provided new discoveries is the calculation of hydride transfer in the enzyme liver alcohol dehydrogenase. In this case, quantum tunneling is important for the hydrogen, as it determines the reaction rate.

Coarse-graining and Reduced Representations

At the other end of the detail scale are coarse-grained and lattice models. Instead of explicitly representing every atom of the system, one uses "pseudo-atoms" to represent groups of atoms. MD simulations on very large systems may require such large computer resources that they cannot easily be studied by traditional all-atom methods. Similarly, simulations of processes on long timescales (beyond about 1 microsecond) are prohibitively expensive, because they require so many time steps. In these cases, one can sometimes tackle the problem by using reduced representations, which are also called coarse-grained models.

Examples for coarse graining (CG) methods are discontinuous molecular dynamics (CG-DMD) and Go-models. Coarse-graining is done sometimes taking larger pseudo-atoms. Such united atom approximations have been used in MD simulations of biological membranes. Implementation of such approach on systems where electrical properties are of interest can be challenging owing to the difficulty of using a proper charge distribution on the pseudo-atoms. The aliphatic tails of lipids are represented by a few pseudo-atoms by gathering 2 to 4 methylene groups into each pseudo-atom.

The parameterization of these very coarse-grained models must be done empirically, by matching the behavior of the model to appropriate experimental data or all-atom simulations. Ideally, these parameters should account for both enthalpic and entropic contributions to free energy in an implicit way. When coarse-graining is done at higher levels, the accuracy of the dynamic description may be less reliable. But very coarse-grained models have been used successfully to examine a wide range of questions in structural biology, liquid crystal organization, and polymer glasses.

Examples of applications of coarse-graining:

- protein folding and protein structure prediction studies are often carried out using one, or a few, pseudo-atoms per amino acid;

- liquid crystal phase transitions have been examined in confined geometries and/or during flow using the Gay-Berne potential, which describes anisotropic species;

- Polymer glasses during deformation have been studied using simple harmonic or FENE springs to connect spheres described by the Lennard-Jones potential;

- DNA supercoiling has been investigated using 1–3 pseudo-atoms per basepair, and at even lower resolution;

- Packaging of double-helical DNA into bacteriophage has been investigated with models where one pseudo-atom represents one turn (about 10 basepairs) of the double helix;

- RNA structure in the ribosome and other large systems has been modeled with one pseudo-atom per nucleotide.

The simplest form of coarse-graining is the *united atom* (sometimes called *extended atom*) and was used in most early MD simulations of proteins, lipids, and nucleic acids. For example, instead of treating all four atoms of a CH_3 methyl group explicitly (or all three atoms of CH_2 methylene group), one represents the whole group with one pseudo-atom. It must, of course, be properly parameterized so that its van der Waals interactions with other groups have the proper distance-dependence. Similar considerations apply to the bonds, angles, and torsions in which the pseudo-atom participates. In this kind of united atom representation, one typically eliminates all explicit hydrogen atoms except those that have the capability to participate in hydrogen bonds (*polar hydrogens*). An example of this is the CHARMM 19 force-field.

The polar hydrogens are usually retained in the model, because proper treatment of hydrogen bonds requires a reasonably accurate description of the directionality and the electrostatic interactions between the donor and acceptor groups. A hydroxyl group, for example, can be both a hydrogen bond donor, and a hydrogen bond acceptor, and it would be impossible to treat this with one OH pseudo-atom. About half the atoms in a protein or nucleic acid are non-polar hydrogens, so the use of united atoms can provide a substantial savings in computer time.

Incorporating Solvent Effects

In many simulations of a solute-solvent system the main focus is on the behavior of the solute with little interest of the solvent behavior particularly in those solvent molecules residing in regions far from the solute molecule. Solvents may influence the dynamic behavior of solutes via random collisions and by imposing a fictional drag on the motion of the solute through the solvent. The use of non-rectangular periodic boundary conditions, stochastic boundaries and solvent shells can all help reduce the amount of solvent molecules required and enable a larger proportion of the computing time to be spend instead on simulating the solute. It is also possible to incorporate the effects of a solvent without needing any explicit solvent molecules present. One example of this approach is to use a potential mean force (PMF) which describes how the free energy changes as a particular coordinate is varied. The free energy change described by PMF contains the averaged effects of the solvent.

Steered Molecular Dynamics (SMD)

Steered molecular dynamics (SMD) simulations, or force probe simulations, apply forces to a protein in order to manipulate its structure by pulling it along desired degrees of freedom. These experiments can be used to reveal structural changes in a protein at the atomic level. SMD is often used to simulate events such as mechanical unfolding or stretching.

There are two typical protocols of SMD: one in which pulling velocity is held constant, and one in which applied force is constant. Typically, part of the studied system (e.g., an atom in a protein) is restrained by a harmonic potential. Forces are then applied to specific atoms at either a constant velocity or a constant force. Umbrella sampling is used to move the system along the desired reaction coordinate by varying, for example, the forces, distances, and angles manipulated in the simulation. Through umbrella sampling, all of the system's configurations—both high-energy and low-energy—are adequately sampled. Then, each configuration's change in free energy can be calculated as the potential of mean force. A popular method of computing PMF is through the weighted histogram analysis method (WHAM), which analyzes a series of umbrella sampling simulations.

Examples of Applications

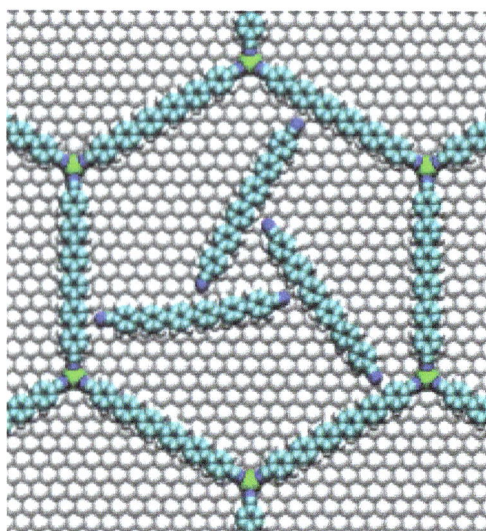

Molecular dynamics simulation of a synthetic molecular motor composed of three molecules in a nanopore (outer diameter 6.7 nm) at 250 K.

Molecular dynamics is used in many fields of science.

- First MD simulation of a simplified biological folding process was published in 1975. Its simulation published in Nature paved the way for the vast area of modern computational protein-folding.

- First MD simulation of a biological process was published in 1976. Its simulation published in Nature paved the way for understanding protein motion as essential in function and not just accessory.

- MD is the standard method to treat collision cascades in the heat spike regime, i.e., the effects that energetic neutron and ion irradiation have on solids and solid surfaces.

- MD simulations were successfully applied to predict the molecular basis of the most common protein mutation N370S, causing Gaucher Disease. In a follow-up publication it was shown that these blind predictions show a surprisingly high correlation with experimental work on the same mutant, published independently at a later point.

- MD simulations have been used to investigate the effect of surface charges on disjoining pressure of thin water films on metal surfaces.

- MD simulations are used along with multislice image simulations to understand transmission electron microscope image features

- MD calculations coupled with Hamiltonian-biasing algorithms have been used to probe the encapsulation thermodynamics of DNA double-strands onto hydrophobic and hydrophilic single-walled carbon nanotubes.

The following biophysical examples illustrate notable efforts to produce simulations of a systems of very large size (a complete virus) or very long simulation times (up to 1.112 milliseconds):

- MD simulation of the full *satellite tobacco mosaic virus* (STMV) (2006, Size: 1 million atoms, Simulation time: 50 ns, program: NAMD) This virus is a small, icosahedral plant virus that worsens the symptoms of infection by Tobacco Mosaic Virus (TMV). Molecular dynamics simulations were used to probe the mechanisms of viral assembly. The entire STMV particle consists of 60 identical copies of one protein that make up the viral capsid (coating), and a 1063 nucleotide single stranded RNA genome. One key finding is that the capsid is very unstable when there is no RNA inside. The simulation would take one 2006 desktop computer around 35 years to complete. It was thus done in many processors in parallel with continuous communication between them.

- Folding simulations of the Villin Headpiece in all-atom detail (2006, Size: 20,000 atoms; Simulation time: 500 μs= 500,000 ns, Program: Folding@

home) This simulation was run in 200,000 CPU's of participating personal computers around the world. These computers had the Folding@home program installed, a large-scale distributed computing effort coordinated by Vijay Pande at Stanford University. The kinetic properties of the Villin Headpiece protein were probed by using many independent, short trajectories run by CPU's without continuous real-time communication. One method employed was the Pfold value analysis, which measures the probability of folding before unfolding of a specific starting conformation. Pfold gives information about transition state structures and an ordering of conformations along the folding pathway. Each trajectory in a Pfold calculation can be relatively short, but many independent trajectories are needed.

- Long continuous-trajectory simulations have been performed on Anton, a massively parallel supercomputer designed and built around custom application-specific integrated circuits (ASICs) and interconnects by D. E. Shaw Research. The longest published result of a simulation performed using Anton is a 1.112-millisecond simulation of NTL9 at 355 K; a second, independent 1.073-millisecond simulation of this configuration was also performed (and many other simulations of over 250 μs continuous chemical time). In *How Fast-Folding Proteins Fold*, researchers Kresten Lindorff-Larsen, Stefano Piana, Ron O. Dror, and David E. Shaw discuss "the results of atomic-level molecular dynamics simulations, over periods ranging between 100 μs and 1 ms, that reveal a set of common principles underlying the folding of 12 structurally diverse proteins." Examination of these diverse long trajectories, enabled by specialized, custom hardware, allow them to conclude that "In most cases, folding follows a single dominant route in which elements of the native structure appear in an order highly correlated with their propensity to form in the unfolded state." In a separate study, Anton was used to conduct a 1.013-millisecond simulation of the native-state dynamics of bovine pancreatic trypsin inhibitor (BPTI) at 300 K.

- These molecular simulations have been used to understand the material removal mechanisms, effects of tool geometry, temperature, and process parameters such as cutting speed and cutting forces. It was also used to investigate the mechanisms behind the exfoliation of few layers of graphene and carbon nanoscrolls.

Molecular Dynamics Algorithms

- Screened Coulomb Potentials Implicit Solvent Model

Integrators

- Symplectic integrator
- Verlet-Stoermer integration

- Runge–Kutta integration
- Beeman's algorithm
- Constraint algorithms (for constrained systems)

Short-range Interaction Algorithms

- Cell lists
- Verlet list
- Bonded interactions

Long-range Interaction Algorithms

- Ewald summation
- Particle mesh Ewald summation (PME)
- Particle–particle-particle–mesh (P3M)
- Shifted force method

Parallelization Strategies

- Domain decomposition method (Distribution of system data for parallel computing)

Specialized Hardware for MD Simulations

- Anton – A specialized, massively parallel supercomputer designed to execute MD simulations
- MDGRAPE – A special purpose system built for molecular dynamics simulations, especially protein structure prediction

Graphics Card as a Hardware for MD Simulations

- Molecular modeling on GPU

QM/MM

The hybrid QM/MM (quantum mechanics/molecular mechanics) approach is a molecular simulation method that combines the strengths of the QM (accuracy) and MM (speed) approaches, thus allowing for the study of chemical processes in solution and in proteins. The QM/MM approach was introduced in the 1976 paper of Warshel and Levitt. They, along with Martin Karplus, won the 2013 Nobel Prize in Chemistry for "the development of multiscale models for complex chemical systems".

An important advantage of QM/MM methods is their efficiency. The cost of doing classical molecular mechanics (MM) simulations in the most straightforward case scales $O(N^2)$, where N is the number of atoms in the system. This is mainly due to electrostatic interactions term (every particle interacts with everything else). However, use of cutoff radius, periodic pair-list updates and more recently the variations of the particle mesh Ewald (PME) method has reduced this to between $O(N)$ to $O(N^2)$. In other words, if a system with twice many atoms is simulated then it would take between twice to four times as much computing power. On the other hand, the simplest *ab-initio* calculations formally scale as $O(N^3)$ or worse (Restricted Hartree–Fock calculations have been suggested to scale $\sim O(N^{2.7})$). Here in the *ab initio* calculations, N stands for the number of basis functions (it is not the number of atoms). Each atom has at least as many basis functions as is the number of electrons (e.g., with the STO-3G basis set). To overcome the limitation, a small part of the system that is of major interest is treated quantum-mechanically (for instance, the active site of an enzyme) and the remaining system is treated classically.

Problems Involved with QM/MM

Even though QM/MM methods are often very efficient, they are still rather tricky to handle. A researcher has to limit the regions (atomic sites) which are simulated by QM. Moving the limitation borders can both effect the results and the time computing the results. The way the QM and MM systems are coupled can differ substantially depending on the arrangement of particles in the system and their deviations from equilibrium positions in time. Usually limits are set at carbon-carbon bonds and avoided in regions that are associated with charged groups, since such an electronically variant limit can influence the quality of the model.

Interatomic Potential

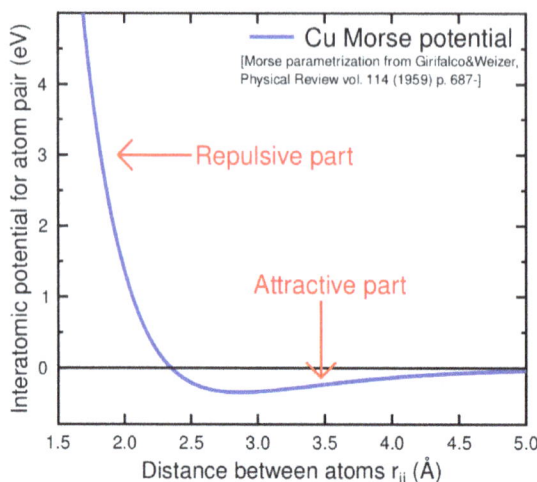

Typical shape of an interatomic pair potential.

Interatomic potentials are mathematical functions for calculating the potential energy of a system of atoms with given positions in space. Interatomic potentials are widely used as the physical basis of molecular mechanics and molecular dynamics simulations in chemistry, molecular physics and materials physics, sometimes in connection with such effects as cohesion, thermal expansion and elastic properties of materials.

Functional Form

Interatomic potentials can be written as a series expansion of functional terms that depend on the position of one, two, three, etc. atoms at a time. Then the total energy of the system V can be written as

$$V_{TOT} = \sum_i^N V_1(\vec{r}_i) + \sum_{i,j}^N V_2(\vec{r}_i, \vec{r}_j) + \sum_{i,j,k}^N V_3(\vec{r}_i, \vec{r}_j, \vec{r}_k) + \cdots$$

Here V_1 is the one-body term, V_2 the two-body term, V_3 the three body term, N the number of atoms in the system, $\vec{r}_i$ the position of atom i, etc. i, j and k are indices that loop over atom positions.

Note that in case the pair potential is given per atom pair, in the two-body term the potential should be multiplied by 1/2 as otherwise each bond is counted twice, and similarly the three-body term by 1/6. Alternatively, the summation of the pair term can be restricted to cases $i < j$ and similarly for the three-body term $i < j < k$, if the potential form is such that it is symmetric with respect to exchange of the j and k indices (this may not be the case for potentials for multielemental systems).

The one-body term is only meaningful if the atoms are in an external field (e.g. an electric field). In the absence of external fields, the potential V should not depend on the absolute position of atoms, but only on the relative positions. This means that the functional form can be rewritten as a function of interatomic distances $r_{ij} = |\vec{r}_i - \vec{r}_j|$ and angles between the bonds (vectors to neighbours) θ_{ijk}. Then, in the absence of external forces, the general form becomes

$$V_{TOT} = \sum_{i,j}^N V_2(r_{ij}) + \sum_{i,j,k}^N V_3(r_{ij}, r_{ik}, \theta_{ijk}) + \cdots$$

In the three-body term V_3 the interatomic distance r_{jk} is not needed since the three terms $r_{ij}, r_{ik}, \theta_{ijk}$ are sufficient to give the relative positions of three atoms i,j,k in three-dimensional space. Any terms of order higher than 2 are also called many-body potentials. In some interatomic potentials the manybody interactions are embedded into the terms of a pair potential .

In principle the sums in the expressions run over all N atoms. However, if the range of the interatomic potential is finite, i.e. the potentials $V(r) \equiv 0$ above some cutoff distance r_{cut}, the summing can be restricted to atoms within the cutoff distance of each other. By also using a cellular method for finding the neighbours, the MD algorithm can be an O(N) algorithm. Potentials with an infinite range can be summed up efficiently by Ewald summation and its further developments.

Force Calculation

The forces acting between atoms can be obtained by differentiation of the total energy with respect to atom positions. That is, to get the force on atom i one should take the three-dimensional derivative (gradient) with respect to the position of atom i:

$$\vec{F}_i = \nabla_{\vec{r}_i} V_{TOT}$$

For two-body potentials this gradient reduces, thanks to the symmetry with respect to ij in the potential form, to straightforward differentiation with respect to the interatomic distances r_{ij}. However, for many-body potentials (three-body, four-body, etc.) the differentiation becomes considerably more complex   since the potential may not be any longer symmetric with respect to ij exchange. In other words, also the energy of atoms k that are not direct neighbours of i can depend on the position $\vec{r}_i$ because of angular and other many-body terms, and hence contribute to the gradient $\nabla_{\vec{r}_i}$.

Classes of Interatomic Potentials

Interatomic potentials come in many different varieties, with different physical motivations. Even for single well-known elements such as silicon, a wide variety of potentials quite different in functional form and motivation have been developed. The true interatomic interactions are quantum mechanical in nature, and there is no known way in which the true interactions described by the Schrödinger equation or Dirac equation for all electrons and nuclei could be cast into an analytical functional form. Hence all analytical interatomic potentials are by necessity approximations.

Pair Potentials

The arguably simplest widely used interatomic interaction model is the Lennard-Jones potential

$$V_{LJ} = 4\varepsilon \left[\left(\frac{\sigma}{r} \right)^{12} - \left(\frac{\sigma}{r} \right)^{6} \right]$$

where ε is the depth of the potential well and σ is the distance at which the potential crosses zero. The term proportional to $1/r^6$ in the potential can be motivated from a classical or quantum mechanical description of the interaction between induced electric dipoles. This potential seems to be quite accurate for noble gases, and is widely used for systems where dipole interactions are significant, including in chemistry force fields to describe intermermolecular interactions.

Another simple and widely used pair potential is the Morse potential, which consists simply of a sum of two exponentials.

$$V(r) = D_e(e^{-2a(r-r_e)} - 2e^{-a(r-r_e)})$$

Here D_e is the equilibrium bond energy and r_e the bond distance. The Morse potential has been applied to studies of molecular vibrations and solids , and although rarely used anymore, inspired the functional form of more modern potentials such as the bond-order potentials.

Ionic materials are often described by a sum of a short-range repulsive term, such as the Buckingham pair potential, and a long-range Coulomb potential giving the ionic interactions between the ions forming the material. The short-range term for ionic materials can also be of many-body character .

Pair potentials have some inherent limitations, like the inability to describe all 3 elastic constants of cubic metals. Hence modern molecular dynamics simulations are to a large extent carried out with different kinds of many-body potentials.

Many-body Potentials

The Stilinger-Weber potential is a potential that has a two-body and three-body terms of the standard form

$$V_{TOT} = \sum_{i,j}^{N} V_2(r_{ij}) + \sum_{i,j,k}^{N} V_3(r_{ij}, r_{ik}, \theta_{ijk})$$

where the three-body term describes how the potential energy changes with bond bending. It was originally developed for pure Si, but has been extended to many other elements and compounds and also formed the basis for other Si potentials.

Metals are very commonly described with what can be called "EAM-like" potentials, i.e. potentials that share the same functional form as the embedded atom model. In these potentials, the total potential energy is written

$$V_{TOT} = \sum_{i}^{N} F_i\left(\sum_{j} \rho(r_{ij})\right) + \frac{1}{2}\sum_{i,j}^{N} V_2(r_{ij})$$

where F_i is a so-called embedding function (different than force $\vec{F}_i$) that is a function of the sum of the so-called electron density $\rho(r_{ij})$. V_2 is a pair potential that usually is purely repulsive. In the original formulation the electron density function $\rho(r_{ij})$ was obtained from true atomic electron densities, and the embedding function was motivated from density-functional theory as the energy needed to 'embed' an atom into the electron density. However, many other potentials used for metals share the same functional form but motivate the terms differently, e.g. based on tight-binding theory or other motivations.

EAM-like potentials are usually implemented as numerical tables. A collection of tables is available at the interatomic potential repository at NIST.

Covalently bonded materials are often described by bond order potentials, sometimes also called Tersoff-like or Brenner-like potentials.

These have in general a form that resembles a pair potential:

$$V_{ij}(r_{ij}) = V_{repulsive}(r_{ij}) + b_{ijk}V_{attractive}(r_{ij})$$

where the repulsive and attractive part are simple exponential functions similar to those in the Morse potential. However, the strength is modified by the environment of the atom i via the b_{ijk} term. If implemented without an explicit angular dependence, these potentials can be shown to be mathematically equivalent to some varieties of EAM-like potentials Thanks to this equivalence, the bond-order potential formalism has been implemented also for many metal-covalent mixed materials.

Repulsive Potentials for Short-range Interactions

For very short interatomic separations, important in radiation material science, the interactions can be described quite accurately with screened Coulomb potentials which have the general form

$$V(r_{ij}) = \frac{1}{4\pi\varepsilon_0} \frac{Z_1 Z_2 e^2}{r_{ij}} \varphi(r/a)$$

here $\varphi(r) \to 1$ when $r \to 0$. Here Z_1 and Z_2 are the charges of the interacting nuclei, and a is the so-called screening parameter. A widely used popular screening function is the "Universal ZBL" one. and more accurate ones can be obtained from all-electron quantum chemistry calculations In binary collision approximation simulations this kind of potential can be used to describe the nuclear stopping power.

Potential Fitting

Since the interatomic potentials are approximations, they by necessity all involve parameters that need to be adjusted to some reference values. In simple potentials such as the Lennard-Jones and Morse ones, the parameters can be set directly to match e.g. the equilibrium bond length and bond strength of a dimer molecule or the cohesive energy of a solid . However, many-body potentials often contain tens or even hundreds of adjustable parameters. These can be fit into a larger set of experimental data, or materials properties derived from more fundamental simulation models such as density-functional theory. For solids, a well-constructed many-body potential can often describe at least the equilibrium crystal structure cohesion and lattice constant, linear elastic constants, and basic point defect properties of all the elements and stable compounds well. The aim of most potential construction and fitting is to make the potential transferable, i.e. that it can describe materials properties that are clearly different from those it was fitted to (for examples of potentials explicitly aiming for this). As an example of demonstrated partial transferability, a review of interatomic potentials of Si found that for instance the Stillinger-Weber and Tersoff III potentials for Si are indeed able to describe several (but certainly not all) materials properties they were not fitted to .

The NIST interatomic potential repository provides a collection of fitted interatomic potentials, either as fitted parameter values or numerical tables of the potential functions.

Reliability of Interatomic Potentials

Classical interatomic potentials cannot reproduce all phenomena. Sometimes quantum description is necessary. Density functional theory is used to overcome this limitation.

Path Integral Molecular Dynamics

Path integral molecular dynamics (PIMD) is a method of incorporating quantum mechanics into molecular dynamics simulations using Feynman path integrals. In PIMD, one uses the Born–Oppenheimer approximation to separate the wavefunction into a nuclear part and an electronic part. The nuclei are treated quantum mechanically by mapping each quantum nucleus onto a classical system of several fictitious particles connected by springs (harmonic potentials) governed by an effective Hamiltonian, which is derived from Feynman's path integral. The resulting classical system, although complex, can be solved relatively quickly. There are now a number of commonly used condensed matter computer simulation techniques that make use of the path integral formulation including Centroid Molecular Dynamics (CMD), Ring Polymer Molecular Dynamics (RPMD), and the Feynman-Kleinert Quasi-Classical Wigner (FK-QCW) method. The same techniques are also used in path integral Monte Carlo (PIMC).

Flying Ice Cube

In molecular dynamics (MD) simulations, the flying ice cube effect is a numerical integration artifact in which the energy of high-frequency fundamental modes is drained into low-frequency modes, particularly into zero-frequency motions such as overall translation and rotation of the system. The artifact derives its name from a particularly noticeable manifestation that arises in simulations of particles in vacuum, where the system being simulated acquires high linear momentum and experiences extremely damped internal motions, freezing the system into a single conformation reminiscent of an ice cube or other rigid body flying through space. The artifact is entirely a consequence of molecular dynamics algorithms and is wholly unphysical, since it violates the principle of equipartition of energy. This is one of several unphysical artifacts that can be observed in molecular dynamics simulations, often arising from the need to balance numerical accuracy with computational efficiency sufficient to achieve adequate sampling of dynamics. The artifact can also occur in generalizations of classical MD simulations, as with Drude oscillators.

The flying ice cube artifact arises from repeated rescalings of the velocities of the particles in the simulation system. The artifact will not occur if the center-of-mass velocity of the system is kept separate and apart from those velocities being rescaled. Velocity rescaling is a means of imposing a thermostat on the system, forcing it to maintain a roughly constant temperature. These rescalings are traditionally done, as in the Berendsen thermostat, by multiplying the system's velocities by a factor α, which equals the ratio of the desired mean kinetic energy divided by the instantaneous amount of kinetic energy. This scheme fails, however, because the instantaneous kinetic energy is located in the *denominator* of the ratio α; fluctuations in the kinetic energy make positive second-order contributions to α, making its average value greater than one even when the instaneous kinetic energy has the proper mean. This causes the constant energy terms — such as those of overall translation and rotation — to grow continuously. Since these energies are constantly increasing, the same rescaling decreases the internal energies, diminishing the internal vibrations. This may be shown mathematically as well; the fluctuating internal kinetic energy has its highs and lows, but its highs are decreased more by velocity rescaling than its lows are increased, leading to a net decrease on average with every rescaling.

When the rotation and translation of the system center of mass are not periodically removed, a particularly noticeable form of the artifact occurs in which nearly all of the system's kinetic energy accrues to these two forms of motion, resulting in a system with essentially no energy associated with internal motions which therefore appears to move as a rigid body. This problem can arise in explicit solvent under unusual circumstances, particularly when the Berendsen barostat is used or when the simulation parameters do not respect conservation of energy, but the artifact occurs most visibly in simulations in vacuum.

Avoidance

The flying ice cube problem in its rigid-body form can be largely avoided by periodically removing the center-of-mass motions, although this does not necessarily cure the less blatant equipartition artifacts. In systems that are simulated as an isolated cluster, such as a single molecule in vacuum, both the translational and rotational motion about the center of mass should be removed; however, for systems in which there is sufficient friction to prevent substantial rotation and many closely spaced fundamental modes between which energy can be transferred - such as those using explicitly represented solvent under periodic boundary conditions - only the translational motion should be removed. Although it does not produce a perfectly continuous trajectory, periodic reassignment of velocities as in the Andersen thermostat method also minimizes the problem.

In complex inhomogeneous systems, such as simulations of membrane proteins in a lipid bilayer, equipartition artifacts are difficult to avoid and may simply be post-processed. Use of Langevin dynamics has been shown to reduce equipartition artifacts in systems containing both structured and unstructured components (i.e., a protein with an intrinsically disordered region).

Metadynamics

Metadynamics (MTD; also abbreviated as METAD or MetaD) is a computer simulation method in computational physics, chemistry and biology. It is used to compute free energy and other state functions of a system, where ergodicity is hindered by the form of the system's energy landscape. It was first suggested by Alessandro Laio and Michele Parrinello in 2002 and is usually applied within molecular dynamics simulations. MTD closely resembles a number of recent methods such as adaptively biased molecular dynamics, adaptive reaction coordinate forces and local elevation umbrella sampling. More recently, both the original and well-tempered metadynamics were derived in the context of importance sampling and shown to be a special case of the adaptive biasing potential setting. MTD is related to the Wang-Landau sampling.

Introduction

The technique builds on a large number of related methods including (in a chronological order) the deflation, tunneling, tabu search, local elevation, conformational flooding, Engkvist-Karlström and adaptive biasing force methods.

Metadynamics has been informally described as "filling the free energy wells with computational sand". The algorithm assumes that the system can be described by a few collective variables. During the simulation, the location of the system in the space determined by the collective variables is calculated and a positive Gaussian potential is added to the real energy landscape of the system. In this way the system is discouraged to come back to the previous point. During the evolution of the simulation, more and more Gaussians sum up, thus discouraging more and more the system to go back to its previous steps, until the system explores the full energy landscape -at this point the modified free energy becomes a constant as a function of the collective variables which is the reason for the collective variables to start fluctuating heavily. At this point the energy landscape can be recovered as the opposite of the sum of all Gaussians.

The time interval between the addition of two Gaussian functions, as well as the Gaussian height and Gaussian width, are tuned to optimize the ratio between accuracy and computational cost. By simply changing the size of the Gaussian, metadynamics can be fitted to yield very quickly a rough map of the energy landscape by using large Gaussians, or can be used for a finer grained description by using smaller Gaussians. Usually, the well-temperated metadynamics is used to change the Gaussian size adaptively. Also, the Gaussian width can be adapted with the adaptive Gaussian metadynamics.

Metadynamics has the advantage, upon methods like adaptive umbrella sampling, of not requiring an initial estimate of the energy landscape to explore. However, it is not trivial to choose proper collective variables for a complex simulation. Typically, it requires several trials to find a good set of collective variables, but there are several automatic procedure proposed: essential coordinates, Sketch-Map, and non-linear data-driven collective variables.

Multi-replica Approach

Independent metadynamics simulations (replicas) can be coupled together to improve usability and parallel performance. There are several such methods proposed: the multiple walker MTD, the parallel tempering MTD, the bias-exchange MTD, and the collective-variable tempering MTD. The last three are similar to the parallel tempering method and use replica exchanges to improve sampling. Typically, the Metropolis–Hastings algorithm is used for replica exchanges, but the infinite swapping and Suwa-Todo algorithms give better replica exchange rates.

Algorithm

Assume, we have a classical N – particle system with positions at $\{\vec{r}_i\}$ $(i \in 1...N)$ in the Cartesian coordinates $(\vec{r}_i \in \mathbb{R}^3)$. The particle interaction are described with a potential function $V \equiv V(\{\vec{r}_i\})$. The potential function form (e.g. two local minima separated by a high-energy barrier) prevents an ergodic sampling with molecular dynamics or Monte Carlo methods.

Original Metadynamics

A general idea of MTD is to enhance the system sampling by discouraging revisiting of sampled states. It is achieved by augmenting the system Hamiltonian H with a bias potential V_{bias} :

$$H = T + V + V_{\text{bias}}.$$

The bias potential is a function of collective variables $(V_{\text{bias}} \equiv V_{\text{bias}}(\vec{s}))$. A collective variable is a function of the particle positions $(\vec{s} \equiv \vec{s}(\{\vec{r}_i\}))$. The bias potential is continuously updated by adding bias at rate ω, where $\vec{s}_t$ is an instantaneous collective variable value at time t:

$$\frac{\partial V_{\text{bias}}(\vec{s})}{\partial t} = \omega \delta(|\vec{s} - \vec{s}_t|).$$

At infinitely long simulation time t_{sim}, the accumulated bias potential converges to free energy with opposite sign (and irrelevant constant C):

$$V_{\text{bias}}(\vec{s}) = \int_0^{t_{\text{sim}}} \omega \delta(|\vec{s} - \vec{s}_t|)\, dt \quad \Rightarrow \quad F(\vec{s}) = -\lim_{t_{\text{sim}} \to \infty} V_{\text{bias}}(\vec{s}) + C$$

For a computationally efficient implementation, the update process is discretised into τ time intervals ($\lfloor\ \rfloor$ denotes the floor function) and δ-function is replaced with a localized positive kernel function K. The bias potential becomes a sum of the kernel functions centred at the instantaneous collective variable values $\vec{s}_j$ at time τj:

$$V_{bias}(\vec{s}) \approx \tau \sum_{j=0}^{\lfloor \frac{t_{\text{sim}}}{\tau} \rfloor} \omega K(|\vec{s} - \vec{s}_j|)$$

.

Typically, the kernel is a multi-dimensional Gaussian function, which covariance matrix has diagonal non-zero elements only:

$$V_{\text{bias}}(\vec{s}) \approx \tau \sum_{j=0}^{\lfloor \frac{t_{\text{sim}}}{\tau} \rfloor} \omega \exp\left(-\frac{1}{2}\left| \frac{\vec{s} - \vec{s}_j}{\vec{\sigma}} \right|^2 \right)$$

.

The parameter τ, ω, and $\vec{\sigma}$ are determined *a priori* and kept constant during the simulation.

Implementation

Below there is a pseudocode of MTD base on molecular dynamics (MD), where $\{\vec{r}\}$ and $\{\vec{v}\}$ are the N-particle system positions and velocities, respectively. The bias V_{bias} is updated every $n = \tau / \Delta t$ MD steps, and its contribution to the system forces $\{\vec{F}\}$ is $\{\vec{F}_{\text{bias}}\}$.

```
set initial {r̄} and {v̄}
set V_bias(s): 0
```

```
every MD step:

    compute CV values:
```

$$\vec{s}_t := \vec{s}(\{\vec{r}\})$$

```
    every n MD steps:
update bias potential:
```

$$V_{\text{bias}}(\vec{s}) := V_{\text{bias}}(\vec{s}) + \tau\omega\exp\left(-\frac{1}{2}\left|\frac{\vec{s}-\vec{s}_t}{\vec{\sigma}}\right|^2\right)$$

```
    compute atomic forces:
```

$$\vec{F}_i := -\frac{\partial V(\{\vec{r}\})}{\partial\vec{r}_i} - \overbrace{\left.\frac{\partial V_{\text{bias}}(\vec{s})}{\partial\vec{s}}\right|_{\vec{s}_t} \frac{\partial\vec{s}(\{\vec{r}\})}{\partial\vec{r}_i}}^{\vec{F}_{\text{bias},i}}$$

```
    propagate {r⃗} and {v⃗} by Δt
```

Free Energy Estimator

The finite size of the kernel makes the bias potential to fluctuate around a mean value. A converged free energy can be obtained by averaging the bias potential. The averaging is started from t_{diff}, when the motion along the collective variable becomes diffusive:

$$\overline{F}(\vec{s}) = -\frac{1}{t_{\text{sim}}-t_{\text{diff}}}\int_{t_{\text{diff}}}^{t_{\text{sim}}}V_{\text{bias}}(\vec{s},t)dt + C$$

Applications

Metadynamics has been used to study:

- protein folding
- chemical reactions
- molecular docking
- phase transitions.
- encapsulation of DNA onto hydrophobic and hydrophilic single-walled carbon nanotubes.

Implementations

Plumed

PLUMED is an open-source library implementing many MTD algorithms and collective variables. It has a flexible object-oriented design and can be interfaced with several MD programs (AMBER, GROMACS, LAMMPS, NAMD, Quantum ESPRESSO, and CP2K).

Other

Other MTD implementations exist in LAMMPS, NAMD, ORAC, CP2K, and Desmond.

Molecular Modelling

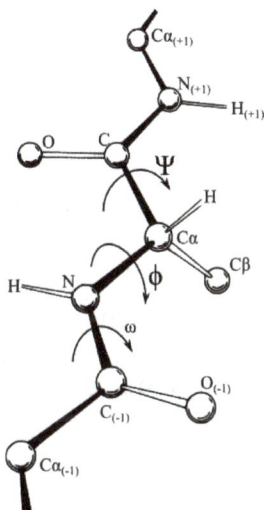

The backbone dihedral angles are included in the molecular model of a protein

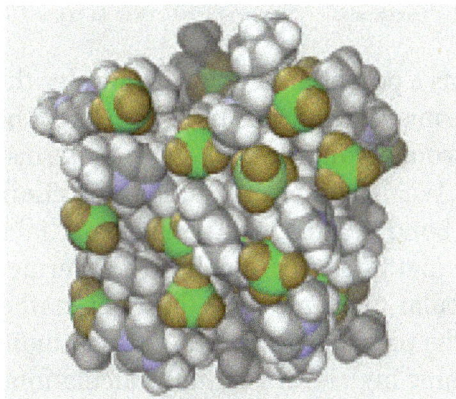

Modeling of ionic liquid

Molecular modelling encompasses all methods, theoretical and computational, used to model or mimic the behaviour of molecules. The methods are used in the fields of computational chemistry, drug design, computational biology and materials science to study molecular systems ranging from small chemical systems to large biological molecules and material assemblies. The simplest calculations can be performed by hand, but inevitably computers are required to perform molecular modelling of any reasonably sized system. The common feature of molecular modelling methods is the atomistic level description of the molecular systems. This may include treating atoms as the smallest individual unit (a molecular mechanics approach), or explicitly modelling electrons of each atom (a quantum chemistry approach).

Molecular Mechanics

Molecular mechanics is one aspect of molecular modelling, as it refers to the use of classical mechanics (Newtonian mechanics) to describe the physical basis behind the models. Molecular models typically describe atoms (nucleus and electrons collectively) as point charges with an associated mass. The interactions between neighbouring atoms are described by spring-like interactions (representing chemical bonds) and Van der Waals forces. The Lennard-Jones potential is commonly used to describe the latter. The electrostatic interactions are computed based on Coulomb's law. Atoms are assigned coordinates in Cartesian space or in internal coordinates, and can also be assigned velocities in dynamical simulations. The atomic velocities are related to the temperature of the system, a macroscopic quantity. The collective mathematical expression is termed a potential function and is related to the system internal energy (U), a thermodynamic quantity equal to the sum of potential and kinetic energies. Methods which minimize the potential energy are termed energy minimization methods (e.g., steepest descent and conjugate gradient), while methods that model the behaviour of the system with propagation of time are termed molecular dynamics.

$$E = E_{\text{bonds}} + E_{\text{angle}} + E_{\text{dihedral}} + E_{\text{non-bonded}}$$

$$E_{\text{non-bonded}} = E_{\text{electrostatic}} + E_{\text{van der Waals}}$$

This function, referred to as a potential function, computes the molecular potential energy as a sum of energy terms that describe the deviation of bond lengths, bond angles and torsion angles away from equilibrium values, plus terms for non-bonded pairs of atoms describing van der Waals and electrostatic interactions. The set of parameters consisting of equilibrium bond lengths, bond angles, partial charge values, force constants and van der Waals parameters are collectively termed a force field. Different implementations of molecular mechanics use different mathematical expressions and different parameters for the potential function. The common force fields in use today have been developed by using high level quantum calculations and/or fitting to experimental data. The method, termed energy minimization, is used to find positions of zero gradient for all atoms, in other words, a local energy minimum. Lower energy states are more stable and are commonly investigated because of their role in chemical and biological processes. A molecular dynamics simulation, on the other hand, computes the behaviour of a system as a function of time. It involves solving Newton's laws of motion, principally the second law, $\mathbf{F} = m\mathbf{a}$. Integration of Newton's laws of motion, using different integration algorithms, leads to atomic trajectories in space and time. The force on an atom is defined as the negative gradient of the potential energy function. The energy minimization method is useful to obtain a static picture for comparing between states of similar systems, while molecular dynamics provides information about the dynamic processes with the intrinsic inclusion of temperature effects.

Variables

Molecules can be modelled either in vacuum, or in the presence of a solvent such as water. Simulations of systems in vacuum are referred to as *gas-phase* simulations, while those that include the presence of solvent molecules are referred to as *explicit solvent* simulations. In another type of simulation, the effect of solvent is estimated using an empirical mathematical expression; these are termed *implicit solvation* simulations.

Coordinate Representations

Most force fields are distance-dependent, making the most convenient expression for these Cartesian coordinates. Yet the comparatively rigid nature of bonds which occur between specific atoms, and in essence, defines what is meant by the designation *molecule*, make an internal coordinate system the most logical representation. In some fields the IC representation is termed the Z-matrix or torsion angle representation. Unfortunately, continuous motions in Cartesian space often require discontinuous angular branches in internal coordinates, making it relatively hard to work with force fields in the internal coordinate representation, and conversely a simple displacement of an atom in Cartesian space may not be a straight line trajectory due to the prohibitions of the interconnected bonds. Thus, it is very common for computational optimizing programs to flip back and forth between representations during their iterations. This can dominate the calculation time of the potential itself and in long chain molecules introduce cumulative numerical inaccuracy. While all conversion algorithms produce mathematically identical results, they differ in speed and numerical accuracy. Currently, the fastest and most accurate torsion to Cartesian conversion is the Natural Extension Reference Frame (NERF) method.

Applications

Molecular modelling methods are now used routinely to investigate the structure, dynamics, surface properties, and thermodynamics of inorganic, biological, and polymeric systems. The types of biological activity that have been investigated using molecular modelling include protein folding, enzyme catalysis, protein stability, conformational changes associated with biomolecular function, and molecular recognition of proteins,

DNA, and membrane complexes.

Molecular Modeling on GPUs

Molecular modeling on GPU is the technique of using a graphics processing unit (GPU) for molecular simulations.

In 2007, NVIDIA introduced video cards that could be used not only to show graphics but also for scientific calculations. These cards include many arithmetic units (as of 2016, up to 3,584 in Tesla P100) working in parallel. Long before this event, the computational power of video cards was purely used to accelerate graphics calculations. What was new is that NVIDIA made it possible to develop parallel programs in a high-level application programming interface (API) named CUDA. This technology substantially simplified programming by enabling programs to be written in C/C++. More recently, OpenCL allows cross-platform GPU acceleration.

Quantum chemistry calculations and molecular mechanics simulations (molecular modeling in terms of classical mechanics) are among beneficial applications of this technology. The video cards can accelerate the calculations tens of times, so a PC with such a card has the power similar to that of a cluster of workstations based on common processors.

GPU Accelerated Molecular Modelling Software

Programs

- Abalone – Molecular Dynamics (Benchmark)
- AMBER on GPUs version
- Ascalaph on GPUs version – Ascalaph Liquid GPU
- BigDFT *Ab initio* program based on wavelet
- Blaze ligand-based virtual screening
- Desmond (software) on GPUs, workstations, and clusters
- Firefly (formerly PC GAMESS)
- FastROCS
- GOMC – GPU Optimized Monte Carlo simulation engine
- GPIUTMD – Graphical processors for Many-Particle Dynamics
- GROMACS on GPUs version

- HALMD – Highly Accelerated Large-scale MD package

- HOOMD-blue – Highly Optimized Object-oriented Many-particle Dynamics—Blue Edition

- LAMMPS on GPUs version – lammps for accelerators

- LIO DFT-Based GPU optimized code -

- Octopus has support for OpenCL.

- oxDNA – DNA and RNA coarse-grained simulations on GPUs

- PWmat – Plane-Wave Density Functional Theory simulations

- TeraChem – Quantum chemistry and *ab initio* Molecular Dynamics

- TINKER on GPUs.

- VMD & NAMD on GPUs versions

- YASARA runs MD simulations on all GPUs using OpenCL.

API

- OpenMM – an API for accelerating molecular dynamics on GPUs, v1.0 provides GPU-accelerated version of GROMACS

- mdcore – an Open-source platform-independent library for molecular dynamics simulations on modern shared-memory parallel architectures.

Distributed Computing Projects

- GPUGRID distributed supercomputing infrastructure

- Folding@home distributed computing project

Monte Carlo Molecular Modeling

Monte Carlo molecular modeling is the application of Monte Carlo methods to molecular problems. These problems can also be modeled by the molecular dynamics method. The difference is that this approach relies on equilibrium statistical mechanics rather than molecular dynamics. Instead of trying to reproduce the dynamics of a system, it generates states according to appropriate Boltzmann probabilities. Thus, it is the application of the Metropolis Monte Carlo simulation to molecular systems. It is therefore also a particular subset of the more general Monte Carlo method in statistical physics.

It employs a Markov chain procedure in order to determine a new state for a system from a previous one. According to its stochastic nature, this new state is accepted at random. Each trial usually counts as a move. The avoidance of dynamics restricts the method to studies of static quantities only, but the freedom to choose moves makes the method very flexible. These moves must only satisfy a basic condition of balance in order equilibrium be properly described, but detailed balance, a stronger condition, is usually imposed when designing new algorithms. An additional advantage is that some systems, such as the Ising model, lack a dynamical description and are only defined by an energy prescription; for these the Monte Carlo approach is the only one feasible.

The great success of this method in statistical mechanics has led to various generalizations such as the method of simulated annealing for optimization, in which a fictitious temperature is introduced and then gradually lowered.

A range of software packages have been developed specifically for the use of the Metropolis Monte Carlo method on molecular simulations. These include:

- BOSS

- MCPro

- Sire

- ProtoMS

Molecular Graphics

Molecular graphics (MG) is the discipline and philosophy of studying molecules and their properties through graphical representation. IUPAC limits the definition to representations on a "graphical display device". Ever since Dalton's atoms and Kekulé's benzene, there has been a rich history of hand-drawn atoms and molecules, and these representations have had an important influence on modern molecular graphics. This topic concentrates on the use of computers to create molecular graphics. Note, however, that many molecular graphics programs and systems have close coupling between the graphics and editing commands or calculations such as in molecular modelling.

Relation to Molecular Models

There has been a long tradition of creating molecular models from physical materials. Perhaps the best known is Crick and Watson's model of DNA built from rods and planar sheets, but the most widely used approach is to represent all atoms and bonds explicitly using the "ball and stick" approach. This can demonstrate a wide range of properties, such as shape, relative size, and flexibility. Many chemistry courses expect that students will have access to ball and stick models. One goal of mainstream molecular graphics has been to represent the "ball and stick" model as realistically as possible and to couple this with calculations of molecular properties.

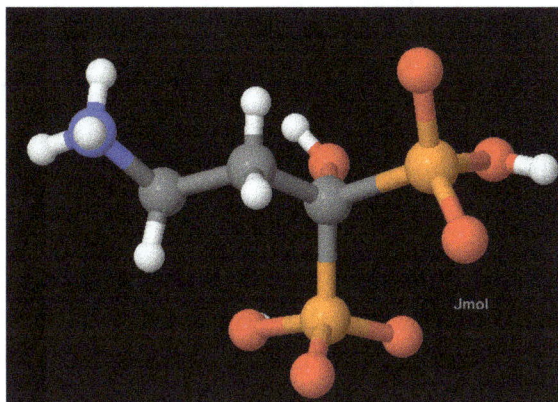

Key: Hydrogen = white, carbon = grey, nitrogen = blue, oxygen = red, and phosphorus = orange

Figure shows a small molecule ($NH_3CH_2CH_2C(OH)(PO_3H)(PO_3H)$-), as drawn by the Jmol program. It is important to realize that the colors and shapes are purely a convention, as individual atoms are not colored, nor do they have hard surfaces. Bonds between atoms are also not rod-shaped.

Comparison of Physical Models with Molecular Graphics

Physical models and computer models have partially complementary strengths and weaknesses. Physical models can be used by those without access to a computer and now can be made cheaply out of plastic materials. Their tactile and visual aspects cannot be easily reproduced by computers (although haptic devices have occasionally been built). On a computer screen, the flexibility of molecules is also difficult to appreciate; illustrating the pseudorotation of cyclohexane is a good example of the value of mechanical models.

However, it is difficult to build large physical molecules, and all-atom physical models of even simple proteins could take weeks or months to build. Moreover, physical models are not robust and they decay over time. Molecular graphics is particularly valuable for representing global and local properties of molecules, such as electrostatic potential. Graphics can also be animated to represent molecular processes and chemical reactions, a feat that is not easy to reproduce physically.

History

Initially the rendering was on early Cathode ray tube screens or through plotters drawing on paper. Molecular structures have always been an attractive choice for developing new computer graphics tools, since the input data are easy to create and the results are usually highly appealing. The first example of MG was a display of a protein molecule (Project MAC, 1966) by Cyrus Levinthal and Robert Langridge. Among the milestones in high-performance MG was the work of Nelson Max in "realistic" rendering of macromolecules using reflecting spheres.

By about 1980 many laboratories both in academia and industry had recognized the power of the computer to analyse and predict the properties of molecules, especially in materials science and the pharmaceutical industry. The discipline was often called "molecular graphics" and in 1982 a group of academics and industrialists in the UK set up the Molecular Graphics Society (MGS). Initially much of the technology concentrated either on high-performance 3D graphics, including interactive rotation or 3D rendering of atoms as spheres (sometimes with radiosity). During the 1980s a number of programs for calculating molecular properties (such as molecular dynamics and quantum mechanics) became available and the term "molecular graphics" often included these. As a result, the MGS has now changed its name to the Molecular Graphics and Modelling Society (MGMS).

The requirements of macromolecular crystallography also drove MG because the traditional techniques of physical model-building could not scale. The first two protein structures solved by molecular graphics without the aid of the Richards' Box were built with Stan Swanson's program FIT on the Vector General graphics display in the laboratory of Edgar Meyer at Texas A&M University: First Marge Legg in Al Cotton's lab at A&M solved the structure of staph. nuclease (1975) and then Jim Hogle solved the structure of monoclinic lysozyme in 1976. A full year passed before other graphics systems were used to replace the Richards' Box for modelling into density in 3-D. Alwyn Jones' FRODO program (and later "O") were developed to overlay the molecular electron density determined from X-ray crystallography and the hypothetical molecular structure.

In 2009 BALLView became the first software to use Raytracing for molecular graphics.

Art, Science and Technology in Molecular Graphics

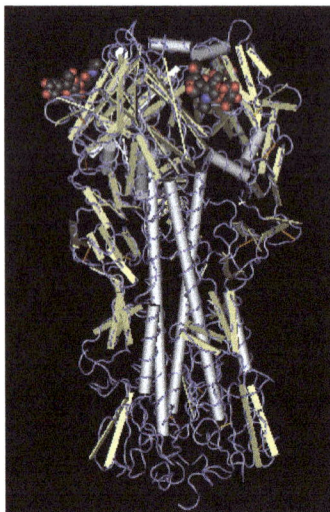

Image of hemagglutinin with alpha helices depicted as cylinders and the rest of the chain as silver coils. The individual protein molecules (several thousand) have been hidden. All of the non-hydrogen atoms in the two ligands (presumably sialic acid) have been shown near the top of the diagram. Key: Carbon = grey, oxygen = red, nitrogen = blue.

Both computer technology and graphic arts have contributed to molecular graphics. The development of structural biology in the 1950s led to a requirement to display molecules with thousands of atoms. The existing computer technology was limited in power, and in any case a naive depiction of all atoms left viewers overwhelmed. Most systems therefore used conventions where information was implicit or stylistic. Two vectors meeting at a point implied an atom or (in macromolecules) a complete residue (10-20 atoms).

The macromolecular approach was popularized by Dickerson and Geis' presentation of proteins and the graphic work of Jane Richardson through high-quality hand-drawn diagrams such as the "ribbon" representation. In this they strove to capture the intrinsic 'meaning' of the molecule. This search for the "messages in the molecule" has always accompanied the increasing power of computer graphics processing. Typically the depiction would concentrate on specific areas of the molecule (such as the active site) and this might have different colors or more detail in the number of explicit atoms or the type of depiction (e.g., spheres for atoms).

In some cases the limitations of technology have led to serendipitous methods for rendering. Most early graphics devices used vector graphics, which meant that rendering spheres and surfaces was impossible. Michael Connolly's program "MS" calculated points on the surface-accessible surface of a molecule, and the points were rendered as dots with good visibility using the new vector graphics technology, such as the Evans and Sutherland PS300 series. Thin sections ("slabs") through the structural display showed very clearly the complementarity of the surfaces for molecules binding to active sites, and the "Connolly surface" became a universal metaphor.

The relationship between the art and science of molecular graphics is shown in the exhibitions sponsored by the Molecular Graphics Society. Some exhibits are created with molecular graphics programs alone, while others are collages, or involve physical materials. An example from Mike Hann (1994), inspired by Magritte's painting *Ceci n'est pas une pipe*, uses an image of a salmeterol molecule. "*Ceci n'est pas une molecule*," writes Mike Hann, "serves to remind us that all of the graphics images presented here are not molecules, not even pictures of molecules, but pictures of icons which we believe represent some aspects of the molecule's properties."

Colour molecular graphics is often use on chemistry journal covers in an artistic manner.

Space-filling Models

Figure is a "space-filling" representation of formic acid, where atoms are drawn as solid spheres to suggest the space they occupy. This and all space-filling models are necessarily icons or abstractions: atoms are nuclei with electron "clouds" of varying density surrounding them, and as such have no actual surfaces. For many years the size of atoms has been approximated by physical models (CPK) in which the volumes of plastic balls describe where much of the electron density is to be found (often sized to van der Waals radii). That is, the surface of these models is meant to represent a specific *level of density* of the electron cloud, not any putative physical surface of the atom.

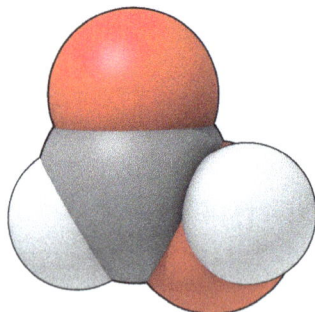

Space-filling model of formic acid. Key: Hydrogen = white, carbon = black, oxygen = red

Since the atomic radii are only slightly less than the distance between bonded atoms, the iconic spheres intersect, and in the CPK models, this was achieved by planar truncations along the bonding directions, the section being circular. When raster graphics became affordable, one of the common approaches was to replicate CPK models *in silico*. It is relatively straightforward to calculate the circles of intersection, but more complex to represent a model with hidden surface removal. A useful side product is that a conventional value for the molecular volume can be calculated.

The use of spheres is often for convenience, being limited both by graphics libraries and the additional effort required to compute complete electronic density or other space-filling quantities. It is now relatively common to see images of surfaces that have been colored to show quantities such as electrostatic potential. Common surfaces in molecular visualization include solvent-accessible ("Lee-Richards") surfaces, solvent-excluded ("Connolly") surfaces, and isosurfaces. The isosurface appears to show the electrostatic potential, with blue colors being negative and red/yellow (near the metal) positive (there is no absolute convention of coloring, and red/positive, blue/negative are often reversed). Opaque isosurfaces do not allow the atoms to be seen and identified and it is not easy to deduce them. Because of this, isosurfaces are often drawn with a degree of transparency.

Technology

Early interactive molecular computer graphics systems were vector graphics machines, which used stroke-writing vector monitors, sometimes even oscilloscopes. The electron beam does not sweep left-and-right as in a raster display. The display hardware followed a sequential list of digital drawing instructions (the display list), directly drawing at an angle one stroke for each molecular bond. When the list was complete, drawing would begin again from the top of the list, so if the list was long (a large number of molecular bonds), the display would flicker heavily. Later vector displays could rotate complex structures with smooth motion, since the orientation of all of the coordinates in the display list could be changed by loading just a few numbers into rotation registers in the display unit, and the display unit would multiply all coordinates in the display list by the contents of these registers as the picture was drawn.

The early black-and white vector displays could somewhat distinguish for example a molecular structure from its surrounding electron density map for crystallographic structure solution work by drawing the molecule brighter than the map. Color display makes them easier to tell apart. During the 1970s two-color stroke-writing Penetron tubes were available, but not used in molecular computer graphics systems. In about 1980 Evans & Sutherland made the first practical full-color vector displays for molecular graphics, typically attached to an E&S PS-300 display. This early color tube was expensive, because it was originally engineered to withstand the shaking of a flight-simulator motion base.

Color raster graphics display of molecular models began around 1978 as seen in this paper by Porter on spherical shading of atomic models. Early raster molecular graphics systems displayed static images that could take around a minute to generate. Dynamically rotating color raster molecular display phased in during 1982-1985 with the introduction of the Ikonas programmable raster display.

Molecular graphics has always pushed the limits of display technology, and has seen a number of cycles of integration and separation of compute-host and display. Early systems like Project MAC were bespoke and unique, but in the 1970s the MMS-X and similar systems used (relatively) low-cost terminals, such as the Tektronix 4014 series, often over dial-up lines to multi-user hosts. The devices could only display static pictures but were able to evangelize MG. In the late 1970s, it was possible for departments (such as crystallography) to afford their own hosts (e.g., PDP-11) and to attach a display (such as Evans & Sutherland's MPS) directly to the bus. The display list was kept on the host, and interactivity was good since updates were rapidly reflected in the display—at the cost of reducing most machines to a single-user system.

In the early 1980s, Evans & Sutherland (E&S) decoupled their PS300 display, which contained its own display information transformable through a dataflow architecture. Complex graphical objects could be downloaded over a serial line (e.g. 9600 baud) and then manipulated without impact on the host. The architecture was excellent for high performance display but very inconvenient for domain-specific calculations, such as electron-density fitting and energy calculations. Many crystallographers and modellers spent arduous months trying to fit such activities into this architecture.

The benefits for MG were considerable, but by the later 1980s, UNIX workstations such as Sun-3 with raster graphics (initially at a resolution of 256 by 256) had started to appear. Computer-assisted drug design in particular required raster graphics for the display of computed properties such as atomic charge and electrostatic potential. Although E&S had a high-end range of raster graphics (primarily aimed at the aerospace industry) they failed to respond to the low-end market challenge where single users, rather than engineering departments, bought workstations. As a result, the market for MG displays passed to Silicon Graphics, coupled with the development of minisuper-computers (e.g., CONVEX and Alliant) which were affordable for well-supported MG laboratories. Silicon Graphics provided a graphics language, IrisGL, which was easier to use and more productive than the PS300 architecture. Commercial companies (e.g., Biosym, Polygen/MSI) ported their code to Silicon Graphics, and by the early 1990s, this was the "industry standard". Dial boxes were often used as control devices.

Stereoscopic displays were developed based on liquid crystal polarized spectacles, and while this had been very expensive on the PS300, it now became a commodity item. A common alternative was to add a polarizable screen to the front of the display and to provide viewers with extremely cheap spectacles with orthogonal polarization for sepa-rate eyes. With projectors such as Barco, it was possible to project stereoscopic display onto special silvered screens and supply an audience of hundreds with spectacles. In this way molecular graphics became universally known within large sectors of chemical and biochemical science, especially in the pharmaceutical industry. Because the back-grounds of many displays were black by default, it was common for modelling sessions and lectures to be held with almost all lighting turned off.

In the last decade almost all of this technology has become commoditized. IrisGL evolved to OpenGL so that molecular graphics can be run on any machine. In 1992, Roger Sayle released his RasMol program into the public domain. RasMol contained a very high-performance molecular renderer that ran on Unix/X Window, and Sayle later ported this to the Windows and Macintosh platforms. The Richardsons developed kinemages and the Mage software, which was also multi-platform. By specifying the chemical MIME type, molecular models could be served over the Internet, so that for the first time MG could be distributed at zero cost regardless of platform. In 1995, Birk-beck College's crystallography department used this to run "Principles of Protein Struc-ture", the first multimedia course on the Internet, which reached 100 to 200 scientists.

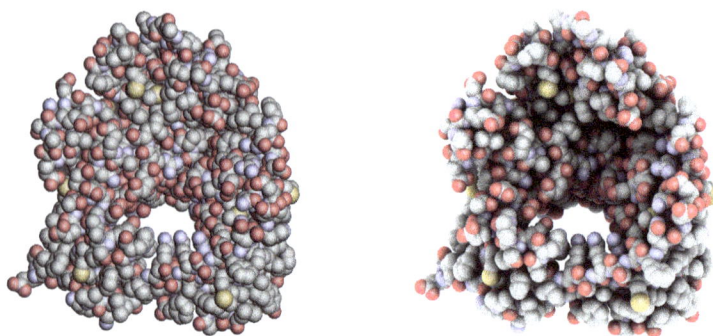

A molecule of Porin (protein) shown without ambient occlusion (left) and with (right). Advanced rendering effects can improve the comprehension of the 3D shape of a molecule.

MG continues to see innovation that balances technology and art, and currently zero-cost or open source programs such as PyMOL and Jmol have very wide use and acceptance.

Recently the widespread diffusion of advanced graphics hardware has improved the rendering capabilities of the visualization tools. The capabilities of current shading languages allow the inclusion of advanced graphic effects (like ambient occlusion, cast shadows and non-photorealistic rendering techniques) in the interactive visualization of molecules. These graphic effects, beside being eye candy, can improve the comprehension of the three-dimensional shapes of the molecules. An example of the effects that can be achieved exploiting recent graphics hardware can be seen in the simple open source visualization system QuteMol.

Algorithms

Reference Frames

Drawing molecules requires a transformation between molecular coordinates (usually, but not always, in Angstrom units) and the screen. Because many molecules are chiral it is essential that the handedness of the system (almost always right-handed) is preserved. In molecular graphics the origin (0, 0) is usually at the lower left, while in many computer systems the origin is at top left. If the z-coordinate is out of the screen (towards the viewer) the molecule will be referred to right-handed axes, while the screen display will be left-handed.

Molecular transformations normally require:

- scaling of the display (but not the molecule).

- translations of the molecule and objects on the screen.

- rotations about points and lines.

Conformational changes (e.g. rotations about bonds) require rotation of one part of the molecule relative to another. The programmer must decide whether a transformation on the screen reflects a change of view or a change in the molecule or its reference frame.

Simple

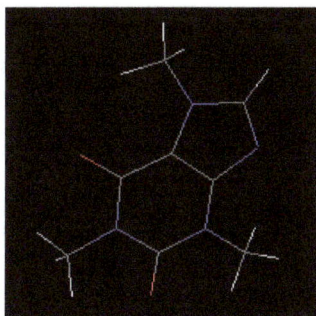

Stick model of caffeine drawn in Jmol

In early displays only vectors could be drawn which are easy to draw because no rendering or hidden surface removal is required.

On vector machines the lines would be smooth but on raster devices Bresenham's algorithm is used (note the "jaggies" on some of the bonds, which can be largely removed with antialiasing softwar).

Atoms can be drawn as circles, but these should be sorted so that those with the largest z-coordinates (nearest the screen) are drawn last. Although imperfect, this often gives a reasonably attractive display. Other simple tricks which do not include hidden surface algorithms are:

- coloring each end of a bond with the same color as the atom to which it is attached.

- drawing less than the whole length of the bond (e.g. 10%-90%) to simulate the bond sticking out of a circle.

- adding a small offset white circle within the circle for an atom to simulate reflection.

Typical pseudocode for creating the figure (to fit the molecule exactly to the screen):

```
// assume:
// atoms with x, y, z coordinates (Angstrom) and elementSymbol
// bonds with pointers/references to atoms at ends
// table of colors for elementTypes
// find limits of molecule in molecule coordinates as xMin, yMin, xMax,
yMax
scale = min(xScreenMax/(xMax-xMin), yScreenMax/(yMax-yMin))
xOffset = -xMin * scale; yOffset = -yMin * scale
for (bond in $bonds) {
  atom0 = bond.getAtom(0)
  atom1 = bond.getAtom(1)
  x0 = xOffset+atom0.getX()*scale; y0 = yOffset+atom0.getY()*scale //
(1)
  x1 = xOffset+atom1.getX()*scale; y1 = yOffset+atom1.getY()*scale //
(2)
  x1 = atom1.getX();   y1 = atom1.getY()
  xMid = (x0 + x1) /2;   yMid = (y0 + y1) /2;
  color0 = ColorTable.getColor(atom0.getSymbol())
```

```
    drawLine (color0, x0, y0, xMid, yMid)

    color1 = ColorTable.getColor(atom1.getSymbol())

    drawLine (color1, x1, y1, xMid, yMid)

}
```

Note that this assumes the origin is in the bottom left corner of the screen, with Y up the screen. Many graphics systems have the origin at the top left, with Y down the screen. In this case the lines (1) and (2) should have the y coordinate generation as:

```
y0 = yScreenMax -(yOffset+atom0.getY()*scale) // (1)

y1 = yScreenMax -(yOffset+atom1.getY()*scale) // (2)
```

Changes of this sort change the handedness of the axes so it is easy to reverse the chirality of the displayed molecule unless care is taken.

Advanced

For greater realism and better comprehension of the 3D structure of a molecule many computer graphics algorithms can be used. For many years molecular graphics has stressed the capabilities of graphics hardware and has required hardware-specific approaches. With the increasing power of machines on the desktop, portability is more important and programs such as Jmol have advanced algorithms that do not rely on hardware. On the other hand, recent graphics hardware is able to interactively render very complex molecule shapes with a quality that would not be possible with standard software techniques.

Electronic Richards Box Systems

Before computer graphics could be employed, mechanical methods were used to fit large molecules to their electron density maps. Using techniques of X-ray crystallography crystal of a substance were bombarded with X-rays, and the diffracted beams that came off were assembled by computer using a Fourier transform into a usually blurry 3-D image of the molecule, made visible by drawing contour circles around high electron density to produce a contoured electron density map.

In the earliest days, contoured electron density maps were hand drawn on large plastic sheets. Sometimes, bingo chips were placed on the plastic sheets where atoms were interpreted to be.

This was superseded by the Richards Box in which an adjustable brass Kendrew molecular model was placed front of a 2-way mirror, behind which were plastic sheets of the electron density map. This optically superimposed the molecular model and the electron density map. The model was moved to within the contour lines of the superimposed map. Then, atomic coordinates were recorded using a plumb bob and a meter stick. Computer graphics held out the hope of vastly speeding up this process, as well as giving a clearer view in many ways.

A noteworthy attempt to overcome the low speed of graphics displays of the time took place at Washington University in St. Louis, USA. Dave Barry's group attempted to leapfrog the state of the art in graphics displays by making custom display hardware to display images complex enough for large-molecule crystallographic structure solution, fitting molecules to their electron-density maps. The MMS-4 display modules were slow and expensive, so a second generation of modules was produced for the MMS-X system.

The first large molecule whose atomic structure was *partly* determined on a molecular computer graphics system was Transfer RNA by Sung-Hou Kim's team in 1976. after initial fitting on a mechanical Richards Box. The first large molecule whose atomic structure was *entirely* determined on a molecular computer graphics system is said to be neurotoxin A from venom of the Philippines sea snake, by Tsernoglou, Petsko, and Tu, with a statement of being first in 1977. The Richardson group published partial atomic structure results of the protein superoxide dismutase the same year, in 1977. All of these were done using the GRIP-75 system.

Other structure fitting systems, FRODO, RING, Builder, MMS-X, etc. succeeded as well within three years and became dominant.

The reason that most of these systems succeeded in just those years, not earlier or later, and within a short timespan had to do with the arrival of commercial hardware that was powerful enough. Two things were needed and arrived at about the same time. First, electron density maps are large and require either a computer with at least a 24-bit address space or a combination of a computer with a lesser 16-bit address space plus several years to overcome the difficulties of an address space that is smaller than the data. The second arrival was that of interactive computer graphics displays that were fast enough to display electron-density maps, whose contour circles require the display of numerous short vectors. The first such displays were the Vector General Series 3 and the Evans and Sutherland Picture System 2, MultiPicture System, and PS-300.

Nowadays, fitting of the molecular structure to the electron density map is largely automated by algorithms with computer graphics a guide to the process. An example is the XtalView XFit program.

Constraint Algorithm

In computational chemistry, a constraint algorithm is a method for satisfying the Newtonian motion of a rigid body which consists of mass points. A restraint algorithm is used to ensure that the distance between mass points is maintained. The general steps involved are; (i) choose novel unconstrained coordinates (internal coordinates), (ii) introduce explicit constraint forces, (iii) minimize constraint forces implicitly by the technique of Lagrange multipliers or projection methods.

Constraint algorithms are often applied to molecular dynamics simulations. Although such simulations are sometimes performed using internal coordinates that automatically satisfy the bond-length, bond-angle and torsion-angle constraints, simulations may also be performed using explicit or implicit constraint forces for these three constraints. However, explicit constraint forces give rise to inefficiency; more computational power is required to get a trajectory of a given length. Therefore, internal coordinates and implicit-force constraint solvers are generally preferred.

Constraint algorithms achieve computational efficiency by neglecting motion along some degrees of freedom. For instance, in atomistic molecular dynamics, typically the length of covalent bonds to hydrogen are constrained; however, constraint algorithms should not be used if vibrations along these degrees of freedom are important for the phenomenon being studied.

Mathematical Background

The motion of a set of N particles can be described by a set of second-order ordinary differential equations, Newton's second law, which can be written in matrix form

$$M \cdot \frac{d^2 \mathbf{q}}{dt^2} = \mathbf{f} = -\frac{\partial V}{\partial \mathbf{q}}$$

where M is a mass matrix and q is the vector of generalized coordinates that describe the particles' positions. For example, the vector q may be a 3N Cartesian coordinates of the particle positions r_k, where k runs from 1 to N; in the absence of constraints, M would be the 3Nx3N diagonal square matrix of the particle masses. The vector f represents the generalized forces and the scalar V(q) represents the potential energy, both of which are functions of the generalized coordinates q.

If M constraints are present, the coordinates must also satisfy M time-independent algebraic equations

$$g_j(\mathbf{q}) = 0$$

where the index j runs from 1 to M. For brevity, these functions g_i are grouped into an M-dimensional vector g below. The task is to solve the combined set of differential-algebraic (DAE) equations, instead of just the ordinary differential equations (ODE) of Newton's second law.

This problem was studied in detail by Joseph Louis Lagrange, who laid out most of the methods for solving it. The simplest approach is to define new generalized coordinates that are unconstrained; this approach eliminates the algebraic equations and reduces the problem once again to solving an ordinary differential equation. Such an approach is used, for example, in describing the motion of a rigid body; the position and orientation of a rigid body can be described by six independent, unconstrained coordinates, rather than describing the positions of the particles that make it up and the constraints among them that maintain their relative distances. The drawback of this approach is that the equations may become unwieldy and complex; for example, the mass matrix M may become non-diagonal and depend on the generalized coordinates.

A second approach is to introduce explicit forces that work to maintain the constraint; for example, one could introduce strong spring forces that enforce the distances among mass points within a "rigid" body. The two difficulties of this approach are that the constraints are not satisfied exactly, and the strong forces may require very short time-steps, making simulations inefficient computationally.

A third approach is to use a method such as Lagrange multipliers or projection to the constraint manifold to determine the coordinate adjustments necessary to satisfy the constraints. Finally, there are various hybrid approaches in which different sets of constraints are satisfied by different methods, e.g., internal coordinates, explicit forces and implicit-force solutions.

Internal Coordinate Methods

The simplest approach to satisfying constraints in energy minimization and molecular dynamics is to represent the mechanical system in so-called *internal coordinates* corresponding to unconstrained independent degrees of freedom of the system. For example, the dihedral angles of a protein are an independent set of coordinates that specify the positions of all the atoms without requiring any constraints. The difficulty of such internal-coordinate approaches is twofold: the Newtonian equations of motion become much more complex and the internal coordinates may be difficult to define for cyclic systems of constraints, e.g., in ring puckering or when a protein has a disulfide bond.

The original methods for efficient recursive energy minimization in internal coordinates were developed by Gō and coworkers.

Efficient recursive, internal-coordinate constraint solvers were extended to molecular dynamics. Analogous methods were applied later to other systems.

Lagrange Multiplier-based Methods

Resolving the constraints of a rigid water molecule using Lagrange multipliers: a) the unconstrained positions are obtained after a simulation time-step, b) the gradients of each constraint over each particle are computed and c) the Lagrange multipliers are computed for each gradient such that the constraints are satisfied.

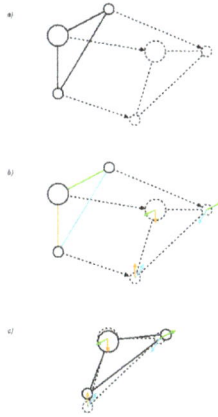

In most of molecular dynamics simulations that use constraint algorithms, constraints are enforced using the method of Lagrange multipliers. Given a set of n linear (holonomic) constraints at the time t,

$$\sigma_k(t) := \| \mathbf{x}_{k\alpha}(t) - \mathbf{x}_{k\beta}(t) \|^2 - d_k^2 = 0, \quad k = 1 \ldots n$$

where $\mathbf{x}_{k\alpha}(t)$ and $\mathbf{x}_{k\beta}(t)$ are the positions of the two particles involved in the kth constraint at the time t and d_k is the prescribed inter-particle distance.

The forces due to these constraints are added in the equations of motion, resulting in, for each of the N particles in the system

$$\frac{\partial^2 \mathbf{x}_i(t)}{\partial t^2} m_i = -\frac{\partial}{\partial \mathbf{x}_i}\left[V(\mathbf{x}_i(t)) + \sum_{k=1}^{n} \lambda_k \sigma_k(t) \right], \quad i = 1 \ldots N.$$

Adding the constraint forces does not change the total energy, as the net work done by the constraint forces (taken over the set of particles that the constraints act on) is zero.

From integrating both sides of the equation with respect to the time, the constrainted coordinates of particles at the time, $t + \Delta t$, are given,

$$(t + \Delta t) = \hat{\mathbf{x}}_i(t + \Delta t) + \sum_{k=1}^{n} \lambda_k \frac{\partial \sigma_k(t)}{\partial \mathbf{x}_i} (\Delta t)^2 m_i^{-1}, \quad i = 1 \ldots N$$

where $\hat{\mathbf{x}}_i(t + \Delta t)$ is the unconstrained (or uncorrected) position of the ith particle after integrating the unconstrained equations of motion.

To satisfy the constraints $\sigma_k(t + \Delta t)$ in the next timestep, the Lagrange multipliers should be determined as the following equation,

$$\sigma_k(t+\Delta t) := \| \mathbf{x}_{k\alpha}(t+\Delta t) - \mathbf{x}_{k\beta}(t+\Delta t) \|^2 - d_k^2 = 0.$$

This implies solving a system of   non-linear equations

$$\sigma_j(t+\Delta t) := \left\| \hat{\mathbf{x}}_{j\alpha}(t+\Delta t) - \hat{\mathbf{x}}_{j\beta}(t+\Delta t) + \sum_{k=1}^{n} \lambda_k (\Delta t)^2 \left[\frac{\partial \sigma_k(t)}{\partial \mathbf{x}_{j\alpha}} m_{j\alpha}^{-1} - \frac{\partial \sigma_k(t)}{\partial \mathbf{x}_{j\beta}} m_{j\beta}^{-1} \right] \right\|^2 - d_j^2 = 0, \quad j=1\ldots n$$

simultaneously for the n unknown Lagrange multipliers λ_k.

This system of n non-linear equations in n unknowns is commonly solved using New-ton–Raphson method where the solution vector $\underline{\lambda}$ is updated using

$$\underline{\lambda}^{(l+1)} \leftarrow \underline{\lambda}^{(l)} - \mathbf{J}_\sigma^{-1} \underline{\sigma}(t+\Delta t)$$

where $\mathbf{J}_\sigma$ is the Jacobian of the equations σ_k:

$$J = \begin{pmatrix} \dfrac{\partial \sigma_1}{\partial \lambda_1} & \dfrac{\partial \sigma_1}{\partial \lambda_2} & \cdots & \dfrac{\partial \sigma_1}{\partial \lambda_n} \\[2mm] \dfrac{\partial \sigma_2}{\partial \lambda_1} & \dfrac{\partial \sigma_2}{\partial \lambda_2} & \cdots & \dfrac{\partial \sigma_2}{\partial \lambda_n} \\[2mm] \vdots & \vdots & \ddots & \vdots \\[2mm] \dfrac{\partial \sigma_n}{\partial \lambda_1} & \dfrac{\partial \sigma_n}{\partial \lambda_2} & \cdots & \dfrac{\partial \sigma_n}{\partial \lambda_n} \end{pmatrix}.$$

Since not all particles contribute to all of constraints, $\mathbf{J}_\sigma$ is a block matrix and can be solved individually to block-unit of the matrix. In other words, $\mathbf{J}_\sigma$ can be solved individually for each molecule.

Instead of constantly updating the vector $\underline{\lambda}$, the iteration can be started with $\underline{\lambda}^{(0)} = \mathbf{0}$, resulting in simpler expressions for $\sigma_k(t)$ and $\dfrac{\partial \sigma_k(t)}{\partial \lambda_j}$. In this case

$$J_{ij} = \frac{\partial \sigma_j}{\partial \lambda_i}\bigg|_{\underline{\lambda}=0} = 2 \left[\hat{x}_{j\alpha} - \hat{x}_{j\beta} \right] \left[\frac{\partial \sigma_i}{\partial x_{j\alpha}} m_{j\alpha}^{-1} - \frac{\partial \sigma_i}{\partial x_{j\beta}} m_{j\beta}^{-1} \right].$$

then λ is updated to

$$\lambda_j = -\mathbf{J}^{-1} \left[\left\| \hat{\mathbf{x}}_{j\alpha}(t+\Delta t) - \hat{\mathbf{x}}_{j\beta}(t+\Delta t) \right\|^2 - d_j^2 \right].$$

After each iteration, the unconstrained particle positions are updated using

$$\hat{\mathbf{x}}_i(t+\Delta t) \leftarrow \hat{\mathbf{x}}_i(t+\Delta t) + \sum_{k=1}^{n} \lambda_k \frac{\partial \sigma_k}{\partial \mathbf{x}_i}.$$

The vector is then reset to

$$\underline{\lambda} = \mathbf{0}.$$

The above procedure is repeated until the solution of constraint equations, $\sigma_k(t+\Delta t)$, converges to a prescribed tolerance of a numerical error.

Although there are a number of algorithms to compute the Lagrange multipliers, these difference is rely only on the methods to solve the system of equations. For this methods, quasi-Newton methods are commonly used.

The SETTLE Algorithm

The SETTLE algorithm solves the system of non-linear equations analytically for $n = 3$ constraints in constant time. Although it does not scale to larger numbers of constraints, it is very often used to constrain rigid water molecules, which are present in almost all biological simulations and are usually modelled using three constraints (e.g. SPC/E and TIP3P water models).

The SHAKE Algorithm

The SHAKE algorithm was first developed for satisfying a bond geometry constraint during molecular dynamics simulations. In SHAKE algorithm, the system of non-linear constraint equations is solved using the Gauss-Seidel method which approximates the solution of the linear system of equations using the Newton-Raphson method;

$$\underline{\lambda} = -\mathbf{J}_\sigma^{-1} \underline{\sigma}.$$

This amounts to assuming that $\mathbf{J}_\sigma$ is diagonally dominant and solving the k th equation only for the k unknown. In practice, we compute

λ_k	$\leftarrow$	$\dfrac{\sigma_k(t)}{\partial \sigma_k(t)/\partial \lambda_k}$,
$\mathbf{x}_{k\alpha}$	$\leftarrow$	$\mathbf{x}_{k\alpha} + \lambda_k \dfrac{\partial \sigma_k(t)}{\partial \mathbf{x}_{k\alpha}}$,
$\mathbf{x}_{k\beta}$	$\leftarrow$	$\mathbf{x}_{k\beta} + \lambda_k \dfrac{\partial \sigma_k(t)}{\partial \mathbf{x}_{k\beta}}$,

for all $k = 1 \ldots n$ iteratively until the constraint equations $\sigma_k(t + \Delta t)$ are solved to a given tolerance.

The calculation cost of each iteration is $\mathcal{O}(n)$, and the iterations themselves converge linearly.

A noniterative form of SHAKE was developed later.

Several variants of the SHAKE algorithm exist. Although they differ in how they compute or apply the constraints themselves, the constraints are still modelled using Lagrange multipliers which are computed using the Gauss-Seidel method.

The original SHAKE algorithm is limited to mechanical systems with a tree structure, i.e., no closed loops of constraints. A later extension of the method, QSHAKE (Quaternion SHAKE) was developed to amend this. It works satisfactorily for *rigid* loops such as aromatic ring systems but fails for flexible loops, such as when a protein has a disulfide bond.

Further extensions include RATTLE, WIGGLE and MSHAKE. RATTLE works the same way as SHAKE, yet using the Velocity Verlet time integration scheme. WIGGLE extends SHAKE and RATTLE by using an initial estimate for the Lagrange multipliers λ_k based on the particle velocities. Finally, MSHAKE computes corrections on the constraint *forces*, achieving better convergence.

A final modification is the P-SHAKE algorithm for rigid or semi-rigid molecules. P-SHAKE computes and updates a pre-conditioner which is applied to the constraint gradients before the SHAKE iteration, causing the Jacobian $\mathbf{J}_\sigma$ to become diagonal or strongly diagonally dominant. The thus de-coupled constraints converge much faster (quadratically as opposed to linearly) at a cost of $\mathcal{O}(n^2)$.

The M-SHAKE Algorithm

The M-SHAKE algorithm solves the non-linear system of equations using Newton's method directly. In each iteration, the linear system of equations

$$\underline{\lambda} = -\mathbf{J}_\sigma^{-1} \underline{\sigma}$$

is solved exactly using an LU decomposition. Each iteration costs $\mathcal{O}(n^3)$ operations, yet the solution converges quadratically, requiring fewer iterations than SHAKE.

This solution was first proposed in 1986 by Ciccotti and Ryckaert under the title "the matrix method", yet differed in the solution of the linear system of equations. Ciccotti and Ryckaert suggest inverting the matrix $\mathbf{J}_\sigma$ directly, yet doing so only once, in the first iteration. The first iteration then costs $\mathcal{O}(n^3)$ operations, whereas the following iterations cost only $\mathcal{O}(n^2)$ operations (for the matrix-vector multiplication). This improvement comes at a cost though, since the Jacobian is no longer updated, convergence is only linear, albeit at a much faster rate than for the SHAKE algorithm.

Several variants of this approach based on sparse matrix techniques were studied by Barth *et al.*

The SHAPE algorithm

The SHAPE algorithm is a multicenter analog of SHAKE for constraining rigid bodies of three or more centers. Like SHAKE, an unconstrained step is taken and then corrected by directly calculating and applying the rigid body rotation matrix that satisfies:

$$L^{rigid}\left(t+\frac{\Delta t}{2}\right) = L^{nonrigid}\left(t+\frac{\Delta t}{2}\right)$$

This approach involves a single 3x3 matrix diagonalization followed by three or four rapid Newton iterations to determine the rotation matrix. SHAPE provides the identical trajectory that is provided with fully converged iterative SHAKE, yet it is found to be more efficient and more accurate than SHAKE when applied to systems involving three or more centers. It extends the ability of SHAKE like constraints to linear systems with three or more atoms, planar systems with four or more atoms, and to significantly larger rigid structures where SHAKE is intractable. It also allows rigid bodies to be linked with one or two common centers (e.g. peptide planes) by solving rigid body constraints iteratively in the same basic manner that SHAKE is used for atoms involving more than one SHAKE constraint.

The LINCS Algorithm

An alternative constraint method, LINCS (Linear Constraint Solver) was developed in 1997 by Hess, Bekker, Berendsen and Fraaije, and was based on the 1986 method of Edberg, Evans and Morriss (EEM), and a modification thereof by Baranyai and Evans (BE).

LINCS applies Lagrange multipliers to the constraint forces and solves for the multipliers by using a series expansion to approximate the inverse of the Jacobian $\mathbf{J}_\sigma$:

$$(\mathbf{I}-\mathbf{J}_\sigma)^{-1} = \mathbf{I}+\mathbf{J}_\sigma+\mathbf{J}_\sigma^2+\mathbf{J}_\sigma^3+\ldots$$

in each step of the Newton iteration. This approximation only works for matrices with Eigenvalues smaller than 1, making the LINCS algorithm suitable only for molecules with low connectivity.

LINCS has been reported to be 3-4 times faster than SHAKE.

Hybrid Methods

Hybrid methods have also been introduced in which the constraints are divided into two groups; the constraints of the first group are solved using internal coordinates whereas those of the second group are solved using constraint forces, e.g., by a Lagrange multiplier or projection method. This approach was pioneered by Lagrange, and result in *Lagrange equations of the mixed type*.

Verlet Integration

Verlet integration is a numerical method used to integrate Newton's equations of motion. It is frequently used to calculate trajectories of particles in molecular dynamics simulations and computer graphics. The algorithm was first used in 1791 by Delambre, and has been rediscovered many times since then, most recently by Loup Verlet in the 1960s for use in molecular dynamics. It was also used by Cowell and Crommelin in 1909 to compute the orbit of Halley's Comet, and by Carl Størmer in 1907 to study the trajectories of electrical particles in a magnetic field (hence it is also called Störmer's method). The Verlet integrator provides good numerical stability, as well as other properties that are important in physical systems such as time-reversibility and preservation of the symplectic form on phase space, at no significant additional computational cost over the simple Euler method.

Basic Störmer–Verlet

For a differential equation of second order of the type $\ddot{\vec{x}}(t) = \vec{A}(\vec{x}(t))$ with initial conditions $\vec{x}(t_0) = \vec{x}_0$ and $\dot{\vec{x}}(t_0) = \vec{v}_0$, an approximate numerical solution $\vec{x}_n \approx \vec{x}(t_n)$ at the times $t_n = t_0 + n\Delta t$ with step size $\Delta t > 0$ can be obtained by the following method:

- set $\vec{x}_1 = \vec{x}_0 + \vec{v}_0 \Delta t + \frac{1}{2} A(\vec{x}_0) \Delta t^2$

- for n=1,2,... iterate

$$\vec{x}_{n+1} = 2\vec{x}_n - \vec{x}_{n-1} + A(\vec{x}_n)\Delta t^2.$$

Equations of Motion

Newton's equation of motion for conservative physical systems is

$$M \ddot{\vec{x}}(t) = F(\vec{x}(t)) = -\nabla V(\vec{x}(t))$$

or individually

$$m_k \ddot{\vec{x}}_k(t) = F_k(\vec{x}(t)) = -\nabla_x V(\vec{x}(t))$$

where

- t is the time,

- $\vec{x}(t) = (\vec{x}_1(t), \ldots, \vec{x}_N(t))$ is the ensemble of the position vector of N objects,

- V is the scalar potential function,

- F is the negative gradient of the potential giving the ensemble of forces on the particles,

- M is the mass matrix, typically diagonal with blocks with mass m_k for every particle.

This equation, for various choices of the potential function V, can be used to describe the evolution of diverse physical systems, from the motion of interacting molecules to the orbit of the planets.

After a transformation to bring the mass to the right side and forgetting the structure of multiple particles, the equation may be simplified to

$$\ddot{\vec{x}}(t) = A(\vec{x}(t))$$

with some suitable vector valued function A representing the position dependent acceleration. Typically, an initial position $\vec{x}(0)$ $\vec{x}$ and an initial velocity $\vec{v}(0) = \dot{\vec{x}}(0) = \vec{v}_0$ are also given.

Verlet Integration (without Velocities)

To discretize and numerically solve this initial value problem, a time step $\Delta t > 0$ is chosen and the sampling point sequence $t_n = n\Delta t$ considered. The task is to construct a sequence of points $\vec{x}_n$ that closely follow the points $\vec{x}(t_n)$ on the trajectory of the exact solution.

Where Euler's method uses the forward difference approximation to the first derivative in differential equations of order one, Verlet Integration can be seen as using the central difference approximation to the second derivative:

$$\frac{\Delta^2 \vec{x}_n}{\Delta t^2} = \frac{\frac{\vec{x}_{n+1} - \vec{x}_n}{\Delta t} - \frac{\vec{x}_n - \vec{x}_{n-1}}{\Delta t}}{\Delta t} = \frac{\vec{x}_{n+1} - 2\vec{x}_n + \vec{x}_{n-1}}{\Delta t^2} = \vec{a}_n = A(\vec{x}_n)$$

Verlet integration in the form used as the *Störmer method* uses this equation to obtain the next position vector from the previous two without using the velocity as

$$\vec{x}_{n+1} = 2\vec{x}_n - \vec{x}_{n-1} + \vec{a}_n \Delta t^2, \qquad \vec{a}_n = A(\vec{x}_n).$$

Discretization Error

The time symmetry inherent in the method reduces the level of local errors introduced into the integration by the discretization by removing all odd degree terms, here the terms in Δt of degree three. The local error is quantified by inserting the exact values $\vec{x}(t_{n-1}), \vec{x}(t_n), \vec{x}(t_{n+1})$ into the iteration and computing the Taylor expansions at time $t = t_n$ of the position vector $\vec{x}(t \pm \Delta t)$ in different time directions.

$$\vec{x}(t + \Delta t) = \vec{x}(t) + \vec{v}(t)\Delta t + \frac{\vec{a}(t)\Delta t^2}{2} + \frac{\vec{b}(t)\Delta t^3}{6} + \mathcal{O}(\Delta t^4)$$

$$\vec{x}(t - \Delta t) = \vec{x}(t) - \vec{v}(t)\Delta t + \frac{\vec{a}(t)\Delta t^2}{2} - \frac{\vec{b}(t)\Delta t^3}{6} + \mathcal{O}(\Delta t^4),$$

where $\vec{x}$ is the position, $\vec{v} = \dot{\vec{x}}$ the velocity, $\vec{a} = \ddot{\vec{x}}$ the acceleration and $\vec{b}$ the jerk (third derivative of the position with respect to the time) t.

Adding these two expansions gives

$$\vec{x}(t + \Delta t) = 2\vec{x}(t) - \vec{x}(t - \Delta t) + \vec{a}(t)\Delta t^2 + \mathcal{O}(\Delta t^4).$$

We can see that the first and third-order terms from the Taylor expansion cancel out, thus making the Verlet integrator an order more accurate than integration by simple Taylor expansion alone.

Caution should be applied to the fact that the acceleration here is computed from the exact solution, $\vec{a}(t) = A(\vec{x}(t))$, whereas in the iteration it is computed at the central iteration point, $\vec{a}_n = A(\vec{x}_n)$. In computing the global error, that is the distance between exact solution and approximation sequence, those two terms do not cancel exactly, influencing the order of the global error.

A Simple Example

To gain insight into the relation of local and global errors it is helpful to examine simple examples where the exact as well as the approximative solution can be expressed in explicit formulas. The standard example for this task is the exponential function.

Consider the linear differential equation $\ddot{x}(t) = w^2 x(t)$ with a constant w. Its exact basis solutions are e^{wt} and e^{-wt}.

The Störmer method applied to this differential equation leads to a linear recurrence relation

$$x_{n+1} - 2x_n + x_{n-1} = h^2 w^2 x_n$$
$$\Leftrightarrow x_{n+1} - 2(1 + \tfrac{1}{2}(wh)^2)x_n + x_{n-1} = 0.$$

It .can be solved by finding the roots of its characteristic polynomial

$$q^2 - 2(1+\tfrac{1}{2}(wh)^2)q + 1 = 0.$$

These are $\quad q_\pm = 1 + \tfrac{1}{2}(wh)^2 \pm wh\sqrt{1+\tfrac{1}{4}(wh)^2}.$

The basis solutions of the linear recurrence are $x_n = q_+^n$ and $x_n = q_-^n$. To compare them with the exact solutions, Taylor expansions are computed.

$$q_+ = 1 + \tfrac{1}{2}(wh)^2 + wh(1+\tfrac{1}{8}(wh)^2 - \tfrac{3}{128}(wh)^4 + \mathcal{O}(h^6))$$
$$= 1 + (wh) + \tfrac{1}{2}(wh)^2 + \tfrac{1}{8}(wh)^3 - \tfrac{3}{128}(wh)^5 + \mathcal{O}(h^7).$$

The quotient of this series with the one of the exponential e^{wh} starts with $1 - \tfrac{1}{24}(wh)^3 + \mathcal{O}(h^5)$, so

$$q_+ = (1 - \tfrac{1}{24}(wh)^3 + \mathcal{O}(h^5))e^{wh}$$
$$= e^{-\frac{1}{24}(wh)^3 + \mathcal{O}(h^5)} e^{wh}.$$

From there it follows that for the first basis solution the error can be computed as

$$x_n = q_+^n = e^{-\frac{1}{24}(wh)^2 wt_n + \mathcal{O}(h^4)} e^{wt_n}$$
$$= e^{wt_n}\left(1 - \tfrac{1}{24}(wh)^2 wt_n + \mathcal{O}(h^4)\right)$$
$$= e^{wt_n} + \mathcal{O}(h^2 t_n e^{wt_n}).$$

That is, although the local discretization error is of order 4, due to the second order of the differential equation the global error is of order 2, with a constant that grows exponentially in time.

Starting the Iteration

Note that at the start of the Verlet-iteration at step $n = 1$, time $t = t_1 = \Delta t$, computing $\vec{x}_2$, one already needs the position vector $\vec{x}_1$ at time $t = t_1$. At first sight this could give problems, because the initial conditions are known only at the initial time $t_0 = 0$. However, from these the acceleration $\vec{a}_0 = A(\vec{x}_0)$ is known, and a suitable approximation for the first time step position can be obtained using the Taylor polynomial of degree two:

$$\vec{x}_1 = \vec{x}_0 + \vec{v}_0 \Delta t + \tfrac{1}{2}\vec{a}_0 \Delta t^2 \approx \vec{x}(\Delta t) + \mathcal{O}(\Delta t^3)$$

The error on the first time step calculation then is of order $\mathcal{O}(\Delta t^3)$. This is not considered a problem because on a simulation over a large number of timesteps, the error on the first timestep is only a negligibly small amount of the total error, which at time t_n is of the order $\mathcal{O}(e^{Lt_n}\Delta t^2)$, both for the distance of the position vectors $\vec{x}_n$ to $\vec{x}(t_n)$ as for the distance of the divided differences $\frac{\vec{x}_{n+1}-\vec{x}_n}{\ddot{A}t}$ to $\frac{\vec{x}(t_{n+1})-\vec{x}(t_n)}{\Delta t}$. Moreover, to obtain this second order global error, the initial error needs to be of at least third order.

Non-constant Time Differences

A disadvantage of the Störmer–Verlet method is that if the time-step (Δt) changes, the method does not approximate the solution to the differential equation. This can be corrected using the formula:

$$\vec{x}_{i+1} = \vec{x}_i + (\vec{x}_i - \vec{x}_{i-1})(\Delta t_i / \Delta t_{i-1}) + \vec{a}\Delta t_i^2$$

A more exact derivation uses the Taylor series (to second order) at t_i for times $t_{i+1} = t_i + \Delta t_i$ and $t_{i-1} = t_i - \Delta t_{i-1}$ to obtain after elimination of $\vec{v}_i$

$$\frac{\vec{x}_{i+1}-\vec{x}_i}{\Delta t_i} + \frac{\vec{x}_{i-1}-\vec{x}_i}{\Delta t_{i-1}} = \vec{a}_i \frac{\Delta t_i + \Delta t_{i-1}}{2}$$

so that the iteration formula becomes

$$\vec{x}_{i+1} = \vec{x}_i + (\vec{x}_i - \vec{x}_{i-1})\frac{\Delta t_i}{\Delta t_{i-1}} + \vec{a}_i \frac{\Delta t_i + \Delta t_{i-1}}{2}\Delta t_i$$

Computing Velocities – Störmer–Verlet Method

The velocities are not explicitly given in the basic Störmer equation, but often they are necessary for the calculation of certain physical quantities like the kinetic energy. This can create technical challenges in molecular dynamics simulations, because kinetic energy and instantaneous temperatures at time t cannot be calculated for a system until the positions are known at time $t + \Delta t$. This deficiency can either be dealt with using the Velocity Verlet algorithm, or estimating the velocity using the position terms and the mean value theorem:

$$\vec{v}(t) = \frac{\vec{x}(t+\Delta t) - \vec{x}(t-\Delta t)}{2\Delta t} + \mathcal{O}(\Delta t^2).$$

Note that this velocity term is a step behind the position term, since this is for the velocity at time t, not $t + \Delta t$, meaning that $\vec{v}_n = \frac{\vec{x}_{n+1}-\vec{x}_{n-1}}{2\Delta t}$ is an order two approximation to $\vec{v}(t_n)$. With the same argument, but halving the time step, $\vec{v}_{n+1/2} = \frac{\vec{x}_{n+1}-\vec{x}_n}{\Delta t}$ is an order two approximation to $\vec{v}(t_{n+1/2})$, with $t_{n+1/2} = t_n + \frac{1}{2}\Delta t$.

One can shorten the interval to approximate the velocity at time $t + \Delta t$ at the cost of accuracy:

$$\vec{v}(t+\Delta t) = \frac{\vec{x}(t+\Delta t) - \vec{x}(t)}{\Delta t} + \mathcal{O}(\Delta t).$$

Velocity Verlet

A related, and more commonly used, algorithm is the Velocity Verlet algorithm, similar to the leapfrog method, except that the velocity and position are calculated at the same value of the time variable (Leapfrog does not, as the name suggests). This uses a similar approach but explicitly incorporates velocity, solving the first-timestep problem in the Basic Verlet algorithm:

$$\vec{x}(t+\Delta t) = \vec{x}(t) + \vec{v}(t)\Delta t + \frac{1}{2}\vec{a}(t)\Delta t^2,$$

$$\vec{v}(t+\Delta t) = \vec{v}(t) + \frac{\vec{a}(t) + \vec{a}(t+\Delta t)}{2}\Delta t.$$

It can be shown that the error on the Velocity Verlet is of the same order as the Basic Verlet. Note that the Velocity algorithm is not necessarily more memory consuming, because it's not necessary to keep track of the velocity at every timestep during the simulation. The standard implementation scheme of this algorithm is:

1. Calculate: $\vec{v}\left(t+\frac{1}{2}\Delta t\right) = \vec{v}(t) + \frac{1}{2}\vec{a}(t)\Delta t.$

2. Calculate: $\vec{x}(t+\Delta t) = \vec{x}(t) + \vec{v}\left(t+\frac{1}{2}\Delta t\right)\Delta t.$

3. Derive $\vec{a}(t+\Delta t)$ from the interaction potential using $\vec{x}(t+\Delta t)$.

4. Calculate: $\vec{v}(t+\Delta t) = \vec{v}\left(t+\frac{1}{2}\Delta t\right) + \frac{1}{2}\vec{a}(t+\Delta t)\Delta t.$

Eliminating the half-step velocity, this algorithm may be shortened to

1. Calculate: $\vec{x}(t+\Delta t) = \vec{x}(t) + \vec{v}(t)\Delta t + \frac{1}{2}\vec{a}(t)\Delta t^2.$

2. Derive $\vec{a}(t+\Delta t)$ from the interaction potential using $\vec{x}(t+\Delta t)$.

3. Calculate: $\vec{v}(t+\Delta t) = \vec{v}(t) + \frac{1}{2}\left(\vec{a}(t) + \vec{a}(t+\Delta t)\right)\Delta t.$

Note, however, that this algorithm assumes that acceleration $\vec{a}(t+\Delta t)$ only depends on position $\vec{x}(t+\Delta t)$, and does not depend on velocity $\vec{v}(t+\Delta t)$.

One might note that the long-term results of Velocity Verlet, and similarly of Leapfrog are one order better than the semi-implicit Euler method. The algorithms are almost identical up to a shifted by half of a timestep in the velocity. This is easily proven by rotating the above loop to start at Step 3 and then noticing that the acceleration term in Step 1 could be eliminated by combining Steps 2 and 4. The only difference is that the midpoint velocity in velocity Verlet is considered the final velocity in semi-implicit Euler method.

The global error of all Euler methods is of order one, whereas the global error of this method is, similar to the midpoint method, of order two. Additionally, if the acceleration indeed results from the forces in a conservative mechanical or Hamiltonian system, the energy of the approximation essentially oscillates around the constant energy of the exactly solved system, with a global error bound again of order one for semi-explicit Euler and order two for Verlet-leapfrog. The same goes for all other converered quantities of the system like linear or angular momentum, that are always preserved or nearly preserved in a symplectic integrator.

The Velocity Verlet method is a special case of the Newmark-beta method with $\beta = 0$ and $\gamma = 1/2$.

Error Terms

The local error in position of the Verlet integrator is $O(\Delta t^4)$ as described above, and the local error in velocity is $O(\Delta t^2)$.

The global error in position, in contrast, is $O(\Delta t^2)$ and the global error in velocity is $O(\Delta t^2)$. These can be derived by noting the following:

$$\text{error}\left(x(t_0 + \Delta t)\right) = O(\Delta t^4)$$

and

$$x(t_0 + 2\Delta t) = 2x(t_0 + \Delta t) - x(t_0) + \Delta t^2 x''(t_0 + \Delta t) + O(\Delta t^4)$$

Therefore:

$$\text{error}(x(t_0 + 2\Delta t)) = 2\text{error}(x(t_0 + \Delta t)) + O(\Delta t^4) = 3O(\Delta t^4)$$

Similarly:

$$\text{error}\left(x(t_0 + 3\Delta t)\right) = 6O(\Delta t^4)$$

$$\text{error}\left(x(t_0 + 4\Delta t)\right) = 10O(\Delta t^4)$$

$$\text{error}\left(x(t_0 + 5\Delta t)\right) = 15O(\Delta t^4)$$

Which can be generalized to (it can be shown by induction, but it is given here without proof):

$$\text{error}\left(x(t_0 + n\Delta t)\right) = \frac{n(n+1)}{2}O(\Delta t^4)$$

If we consider the global error in position between $x(t)$ and $x(t+T)$, where $T = n\Delta t$, it is clear that:

$$\text{error}\big(x(t_0 + T)\big) = \left(\frac{T^2}{2\Delta t^2} + \frac{T}{2\Delta t}\right)O(\Delta t^4)$$

And therefore, the global (cumulative) error over a constant interval of time is given by:

$$\text{error}\big(x(t_0 + T)\big) = O(\Delta t^2)$$

Because the velocity is determined in a non-cumulative way from the positions in the Verlet integrator, the global error in velocity is also $O(\Delta t^2)$.

In molecular dynamics simulations, the global error is typically far more important than the local error, and the Verlet integrator is therefore known as a second-order integrator.

Constraints

Systems of multiple particles with constraints are simpler to solve with Verlet integration than with Euler methods. Constraints between points may be for example potentials constraining them to a specific distance or attractive forces. They may be modeled as springs connecting the particles. Using springs of infinite stiffness, the model may then be solved with a Verlet algorithm.

In one dimension, the relationship between the unconstrained positions $\tilde{x}_i^{(t)}$ and the actual positions $x_i^{(t)}$ of points i at time t can be found with the algorithm

$$d_1 = x_2^{(t)} - x_1^{(t)}$$

$$d_2 = \| d_1 \|$$

$$d_3 = \frac{d_2 - r}{d_2}$$

$$x_1^{(t+\Delta t)} = \tilde{x}_1^{(t+\Delta t)} + \frac{1}{2}d_1 d_3$$

$$x_2^{(t+\Delta t)} = \tilde{x}_2^{(t+\Delta t)} - \frac{1}{2}d_1 d_3$$

Verlet integration is useful because it directly relates the force to the position, rather than solving the problem via velocities.

Problems, however, arise when multiple constraining forces act on each particle. One way to solve this is to loop through every point in a simulation, so that at every point the constraint relaxation of the last is already used to speed up the spread of the information. In a simulation this may be implemented by using small time steps for the simulation, using a fixed number of constraint solving steps per time step, or solving constraints until they are met by a specific deviation.

When approximating the constraints locally to first order this is the same as the Gauss–Seidel method. For small matrices it is known that LU decomposition is faster. Large systems can be divided into clusters (for example: each ragdoll = cluster). Inside clusters the LU method is used, between clusters the Gauss–Seidel method is used. The matrix code can be reused: The dependency of the forces on the positions can be approximated locally to first order, and the Verlet integration can be made more implicit.

Sophisticated software, such as SuperLU exists to solve complex problems using sparse matrices. Specific techniques, such as using (clusters of) matrices, may be used to address the specific problem, such as that of force propagating through a sheet of cloth without forming a sound wave.

Another way to solve holonomic constraints is to use constraint algorithms.

Collision Reactions

One way of reacting to collisions is to use a penalty-based system which basically applies a set force to a point upon contact. The problem with this is that it is very difficult to choose the force imparted. Use too strong a force and objects will become unstable, too weak and the objects will penetrate each other. Another way is to use projection collision reactions which takes the offending point and attempts to move it the shortest distance possible to move it out of the other object.

The Verlet integration would automatically handle the velocity imparted via the collision in the latter case, however note that this is not guaranteed to do so in a way that is consistent with collision physics (that is, changes in momentum are not guaranteed to be realistic). Instead of implicitly changing the velocity term, one would need to explicitly control the final velocities of the objects colliding (by changing the recorded position from the previous time step).

The two simplest methods for deciding on a new velocity are perfectly elastic collisions and inelastic collisions. A slightly more complicated strategy that offers more control would involve using the coefficient of restitution.

Implicit Solvation

Implicit solvation (sometimes termed continuum solvation) is a method to represent solvent as a continuous medium instead of individual "explicit" solvent molecules, most often used in molecular dynamics simulations and in other applications of molecular mechanics. The method is often applied to estimate free energy of solute-solvent interactions in structural and chemical processes, such as folding or conformational transitions of proteins, DNA, RNA, and polysaccharides, association of biological macromolecules with ligands, or transport of drugs across biological membranes.

The implicit solvation model is justified in liquids, where the potential of mean force can be applied to approximate the averaged behavior of many highly dynamic solvent molecules. However, the interfaces and the interiors of biological membranes or proteins can also be considered as media with specific solvation or dielectric properties. These media are not necessarily uniform, since their properties can be described by different analytical functions, such as "polarity profiles" of lipid bilayers. There are two basic types of implicit solvent methods: models based on accessible surface areas (ASA) that were historically the first, and more recent continuum electrostatics models, although various modifications and combinations of the different methods are possible. The accessible surface area (ASA) method is based on experimental linear relations between Gibbs free energy of transfer and the surface area of a solute molecule. This method operates directly with free energy of solvation, unlike molecular mechanics or electrostatic methods that include only the enthalpic component of free energy. The continuum representation of solvent also significantly improves the computational speed and reduces errors in statistical averaging that arise from incomplete sampling of solvent conformations, so that the energy landscapes obtained with implicit and explicit solvent are different. Although the implicit solvent model is useful for simulations of biomolecules, this is an approximate method with certain limitations and problems related to parameterization and treatment of ionization effects.

Accessible Surface Area-based Method

The free energy of solvation of a solute molecule in the simplest ASA-based method is given by:

$$\Delta G_{\text{solv}} = \sum_i \sigma_i \, ASA_i$$

where ASA_i is the accessible surface area of atom i, and σ_i is *solvation parameter* of atom i, i.e., a contribution to the free energy of solvation of the particular atom i per surface unit area. The needed solvation parameters for different types of atoms (carbon (C), nitrogen (N), oxygen (O), sulfur (S), etc.) are usually determined by a least squares fit of the calculated and experimental transfer free energies for a series of organic compounds. The experimental energies are determined from partition coefficients of these compounds between different solutions or media using standard mole concentrations of the solutes.

Notably, *solvation energy* is the free energy needed to transfer a solute molecule from a solvent to *vacuum* (gas phase). This energy can supplement the intramolecular energy in vacuum calculated in molecular mechanics. Thus, the needed atomic solvation parameters were initially derived from water-gas partition data. However, the dielectric properties of proteins and lipid bilayers are much more similar to those of nonpolar solvents than to vacuum. Newer parameters have thus been derived from water-1-octanol partition coefficients or other similar data. Such parameters actually describe *transfer* energy between two condensed media or the *difference* of two solvation energies.

Poisson-Boltzmann

Although this equation has solid theoretical justification, it is computationally expensive to calculate without approximations. The Poisson-Boltzmann equation (PB) describes the electrostatic environment of a solute in a solvent containing ions. It can be written in cgs units as:

$$\vec{\nabla} \cdot \left[\epsilon(\vec{r}) \vec{\nabla} \Psi(\vec{r}) \right] = -4\pi \rho^f(\vec{r}) - 4\pi \sum_i c_i^\infty z_i q \lambda(\vec{r}) e^{\frac{-z_i q \Psi(\vec{r})}{kT}}$$

or (in mks):

$$\vec{\nabla} \cdot \left[\epsilon(\vec{r}) \vec{\nabla} \Psi(\vec{r}) \right] = -\rho(\vec{r}) - \sum c_i^\infty z_i q \lambda(\vec{r}) e^{\frac{-z q \Psi(r)}{kT}}$$

where $\epsilon(\vec{r})$ represents the position-dependent dielectric, $\Psi(\vec{r})$ represents the electrostatic potential, $\rho^f(\vec{r})$ represents the charge density of the solute, c_i^∞ represents the concentration of the ion i at a distance of infinity from the solute, z_i is the valence of the ion, q is the charge of a proton, k is the Boltzmann constant, T is the temperature, and $\lambda(\vec{r})$ is a factor for the position-dependent accessibility of position r to the ions in solution (often set to uniformly 1). If the potential is not large, the equation can be linearized to be solved more efficiently.

A number of numerical Poisson-Boltzmann equation solvers of varying generality and efficiency have been developed, including one application with a specialized computer hardware platform. However, performance from PB solvers does not yet equal that from the more commonly used generalized Born approximation.

Generalized Born

The *Generalized Born* (GB) model is an approximation to the exact (linearized) Poisson-Boltzmann equation. It is based on modeling the solute as a set of spheres whose internal dielectric constant differs from the external solvent. The model has the following functional form:

$$G_s = \frac{1}{8\pi\epsilon_0} \left(1 - \frac{1}{\epsilon} \right) \sum_{i,j}^N \frac{q_i q_j}{f_{GB}}$$

where

$$f_{GB} = \sqrt{r_{ij}^2 + a_{ij}^2 e^{-D}}$$

and $D = \left(\dfrac{r_{ij}}{2a_{ij}}\right)^2$, $a_{ij} = \sqrt{a_i a_j}$

where ϵ_0 is the permittivity of free space, ϵ is the dielectric constant of the solvent being modeled, q_i is the electrostatic charge on particle i, r_{ij} is the distance between particles i and j, and a_i is a quantity (with the dimension of length) termed the *effective Born radius*. The effective Born radius of an atom characterizes its degree of burial inside the solute; qualitatively it can be thought of as the distance from the atom to the molecular surface. Accurate estimation of the effective Born radii is critical for the GB model.

With Accessible Surface Area

The Generalized Born (GB) model augmented with the hydrophobic solvent accessible surface area (SA) term is GBSA. It is among the most commonly used implicit solvent model combinations. The use of this model in the context of molecular mechanics is termed MM/GBSA. Although this formulation has been shown to successfully identify the native states of short peptides with well-defined tertiary structure, the conformational ensembles produced by GBSA models in other studies differ significantly from those produced by explicit solvent and do not identify the protein's native state. In particular, salt bridges are overstabilized, possibly due to insufficient electrostatic screening, and a higher-than-native alpha helix population was observed. Variants of the GB model have also been developed to approximate the electrostatic environment of membranes, which have had some success in folding the transmembrane helixes of integral membrane proteins.

Ad hoc Fast Solvation Models

Another possibility is to use ad hoc quick strategies to estimate solvation free energy. A first generation of fast implicit solvents is based on the calculation of a per-atom solvent accessible surface area. For each of group of atom types, a different parameter scales its contribution to solvation.

Another strategy is implemented for the CHARMM19 force-field and is called EEF1. EEF1 is based on a Gaussian-shaped solvent exclusion. The solvation free energy is

$$\Delta G_i^{solv} = \Delta G_i^{ref} - \sum_j \int_{Vj} f_i(r)dr$$

The reference solvation free energy of i corresponds to a suitably chosen small molecule in which group i is essentially fully solvent-exposed. The integral is over the volume V_j of group j and the summation is over all groups j around i. EEF1 additionally uses a distance-dependent (non-constant) dielectric, and ionic side-chains of proteins are simply neutralized. It is only 50% slower than a vacuum simulation. This model was later augmented with the hydrophobic effect and called Charmm19/SASA.

Hybrid Implicit-explicit Solvation Models

It is possible to include a layer or sphere of water molecules around the solute, and model the bulk with an implicit solvent. Such an approach is proposed by M. J. Frisch and co-workers and by other authors. For instance in Ref. the bulk solvent is modeled with a Generalized Born approach and the multi-grid method used for Coulombic pairwise particle interactions. It is reported to be faster than a full explicit solvent simulation with the particle mesh Ewald summation (PME) method of electrostatic calculation. There are a range of hybrid methods available capable of accessing and acquiring information on solvation.

Effects Unaccounted for

The Hydrophobic Effect

Models like PB and GB allow estimation of the mean electrostatic free energy but do not account for the (mostly) entropic effects arising from solute-imposed constraints on the organization of the water or solvent molecules. This is termed the hydrophobic effect and is a major factor in the folding process of globular proteins with hydrophobic cores. Implicit solvation models may be augmented with a term that accounts for the hydrophobic effect. The most popular way to do this is by taking the solvent accessible surface area (SASA) as a proxy of the extent of the hydrophobic effect. Most authors place the extent of this effect between 5 and 45 cal/(Å^2 mol). Note that this surface area pertains to the solute, while the hydrophobic effect is mostly entropic in nature at physiological temperatures and occurs on the side of the solvent.

Viscosity

Implicit solvent models such as PB, GB, and SASA lack the viscosity that water molecules impart by randomly colliding and impeding the motion of solutes through their van der Waals repulsion. In many cases, this is desirable because it makes sampling of configurations and phase space much faster. This acceleration means that more configurations are visited per simulated time unit, on top of whatever CPU acceleration is achieved in comparison to explicit solvent. It can, however, lead to misleading results when kinetics are of interest.

Viscosity may be added back by using Langevin dynamics instead of Hamiltonian mechanics and choosing an appropriate damping constant for the particular solvent. In practical bimolecular simulations one can often speed-up conformational search significantly (up to 100 times in some cases) by using much lower collision frequency γ. Recent work has also been done developing thermostats based on fluctuating hydrodynamics to account for momentum transfer through the solvent and related thermal fluctuations. One should keep in mind, though, that the folding rate of proteins does not depend linearly on viscosity for all regimes.

Hydrogen Bonds with Solvent

Solute-solvent hydrogen bonds in the first solvation shell are important for solubility of organic molecules and especially ions. Their average energetic contribution can be reproduced with an implicit solvent model.

Problems and Limitations

All implicit solvation models rest on the simple idea that nonpolar atoms of a solute tend to cluster together or occupy nonpolar media, whereas polar and charged groups of the solute tend to remain in water. However, it is important to properly balance the opposite energy contributions from different types of atoms. Several important points have been discussed and investigated over the years.

Choice of Model Solvent

It has been noted that wet 1-octanol solution is a poor approximation of proteins or biological membranes because it contains ~2M of water, and that cyclohexane would be a much better approximation. Investigation of passive permeability barriers for different compounds across lipid bilayers led to conclusion that 1,9-decadiene can serve as a good approximations of the bilayer interior, whereas 1-octanol was a very poor approximation. A set of solvation parameters derived for protein interior from protein engineering data was also different from octanol scale: it was close to cyclohexane scale for nonpolar atoms but intermediate between cyclohexane and octanol scales for polar atoms. Thus, different atomic solvation parameters should be applied for modeling of protein folding and protein-membrane binding. This issue remains controversial. The original idea of the method was to derive all solvation parameters directly from experimental partition coefficients of organic molecules, which allows calculation of solvation free energy. However, some of the recently developed electrostatic models use *ad hoc* values of 20 or 40 cal/($Å^2$ mol) for *all* types of atoms. The non-existent "hydrophobic" interactions of polar atoms are overridden by large electrostatic energy penalties in such models.

Solid-state Applications

Strictly speaking, ASA-based models should only be applied to describe *solvation*, i.e., energetics of transfer between liquid or uniform media. It is possible to express van der Waals interaction energies in the solid state in the surface energy units. This was sometimes done for interpreting protein engineering and ligand binding energetics, which leads to "solvation" parameter for aliphatic carbon of ~40 cal/($Å^2$ mol), which is 2 times bigger than ~20 cal/($Å^2$ mol) obtained for transfer from water to liquid hydrocarbons, because the parameters derived by such fitting represent sum of the hydrophobic energy (i.e., 20 cal/Å² mol) and energy of van der Waals attractions of aliphatic groups in the solid state, which corresponds to fusion enthalpy of alkanes. Unfortunately, the simplified ASA-based model cannot capture the "specific" distance-dependent interactions between different types of atoms in the solid state which are responsible for clustering of atoms with similar polarities in protein structures and molecular crystals. Parameters of such interatomic interactions, together with atomic solvation parameters for the protein interior, have been approximately derived from protein engineering data. The implicit solvation model breaks down when solvent molecules associate strongly with binding cavities in a protein, so that the protein and the solvent molecules form a continuous solid body. On the other hand, this model can be successfully applied for describing transfer from water to the *fluid* lipid bilayer.

Importance of Extensive Testing

More testing is needed to evaluate the performance of different implicit solvation models and parameter sets. They are often tested only for a small set of molecules with very simple structure, such as hydrophobic and amphiphilic alpha helixes (α). This method was rarely tested for hundreds of protein structures.

Treatment of Ionization Effects

Ionization of charged groups has been neglected in continuum electrostatic models of implicit solvation, as well as in standard molecular mechanics and molecular dynamics. The transfer of an ion from water to a nonpolar medium with dielectric constant of ~3 (lipid bilayer) or 4 to 10 (interior of proteins) costs significant energy, as follows from the Born equation and from experiments. However, since the charged protein residues are ionizable, they simply lose their charges in the nonpolar environment, which costs relatively little at the neutral pH: ~4 to 7 kcal/mol for Asp, Glu, Lys, and Arg amino acid residues, according to the Henderson-Hasselbalch equation, $\Delta G = 2.3RT\ (pH - pK)$. The low energetic costs of such ionization effects have indeed been observed for protein mutants with buried ionizable residues. and hydrophobic α-helical peptides in membranes with a single ionizable residue in the middle. However, all electrostatic methods, such as PB, GB, or GBSA assume that ionizable groups remain charged in the nonpolar environments, which leads to grossly overestimated electrostatic energy. In the simplest accessible surface area-based models, this problem was treated using different solvation parameters for charged atoms or Henderson-Hasselbalch equation with some modifications. However even the latter approach does not solve the problem. Charged residues can remain charged even in the nonpolar environment if they are involved in intramolecular ion pairs and H-bonds. Thus, the energetic penalties can be overestimated even using the Henderson-Hasselbalch equation. More rigorous theoretical methods describing such ionization effects have been developed, and there are ongoing efforts to incorporate such methods into the implicit solvation models.

Combining Rules

In computational chemistry and molecular dynamics, the combination rules or combining rules are equations that provide the interaction energy between two dissimilar non-bonded atoms, usually for the part of the potential representing the van der Waals interaction. In the simulation of mixtures, the choice of combining rules can sometimes affect the outcome of the simulation.

Combining Rules for the Lennard-Jones Potential

The Lennard-Jones Potential is a mathematically simple model for the interaction between a pair of atoms or molecules. One of the most common forms is

$$V_{LJ} = 4\varepsilon \left[\left(\frac{\sigma}{r} \right)^{12} - \left(\frac{\sigma}{r} \right)^{6} \right]$$

where ε is the depth of the potential well, σ is the finite distance at which the inter-particle potential is zero, r is the distance between the particles. The potential reaches a minimum, of depth ε, when $r = 2^{1/6}\sigma \approx 1.122\sigma$.

Lorentz-Berthelot Rules

The Lorentz rule was proposed by H. A. Lorentz in 1881:

$$\sigma_{ij} = \frac{\sigma_{ii} + \sigma_{jj}}{2}$$

The Lorentz rule is only analytically correct for hard sphere systems.

The Berthelot rule (Daniel Berthelot, 1898) is given by:

$$\epsilon_{ij} = \sqrt{\epsilon_{ii}\epsilon_{jj}}$$

These rules are the most widely used and are the default in many molecular simulation packages, but are not without failings.

Waldman-Hagler Rules

The Waldman-Hagler rules are given by:

$$r_{ij}^{0} = \left(\frac{(r_i^0)^6 + (r_j^0)^6}{2} \right)^{1/6}$$

and

$$\epsilon_{ij} = 2\sqrt{\epsilon_i \cdot \epsilon_j} \left(\frac{(r_i^0)^3 \cdot (r_j^0)^3}{(r_i^0)^6 + (r_j^0)^6} \right)$$

Fender-Halsey

The Fender-Halsey combining rule is given by

$$\epsilon_{ij} = \frac{2\epsilon_i \epsilon_j}{\epsilon_i + \epsilon_j}$$

Kong Rules

The Kong rules for the Lennard-Jones potential are given by:

$$\epsilon_{ij}\sigma_{ij}^6 = \left(\epsilon_{ii}\sigma_{ii}^6 \grave{o}_{jj}\sigma_{jj}^6 \right)^{1/2}$$

$$\epsilon_{ij}\sigma_{ij}^{12} = \left[\frac{(\epsilon_{ii}\sigma_{ii}^{12})^{1/13} + (\epsilon_{jj}\sigma_{jj}^{12})^{1/13}}{2} \right]^{13}$$

Others

Many others have been proposed, including those of Tang and Toennies Pena, Hudson and McCoubrey and Sikora(1970).

Combining Rules for other Potentials

Good-Hope Rule

The Good-Hope rule for Mie–Lennard-Jones or Buckingham potentials is given by:

$$\sigma_{ij} = \sqrt{\sigma_{ii}\sigma_{jj}}$$

Hogervorst Rules

The Hogervorst rules for the Exp-6 potential are:

$$\epsilon_{12} = \frac{2\epsilon_{11}\epsilon_{22}}{\epsilon_{11} + \epsilon_{22}}$$

and

$$\alpha_{12} = \frac{1}{2}(\alpha_{11} + \alpha_{22})$$

Kong-Chakrabarty Rules

The Kong-Chakrabarty rules for the Exp-6 potential are:

$$\left[\frac{\epsilon_{12}\alpha_{12}e^{\alpha_{12}}}{(\alpha_{12}-6)\sigma_{12}}\right]^{2\sigma_{12}/\alpha_{12}} = \left[\frac{\epsilon_{11}\alpha_{11}e^{\alpha_{11}}}{(\alpha_{11}-6)\sigma_{11}}\right]^{\sigma_{11}/\alpha_{11}}\left[\frac{\epsilon_{22}\alpha_{22}e^{\alpha_{22}}}{(\alpha_{22}-6)\sigma_{22}}\right]^{\sigma_{22}/\alpha_{22}}$$

$$\frac{\sigma_{12}}{\alpha_{12}} = \frac{1}{2}\left(\frac{\sigma_{11}}{\alpha_{11}}+\frac{\sigma_{22}}{\alpha_{22}}\right)$$

and

$$\frac{\epsilon_{12}\alpha_{12}\sigma_{12}^6}{(\alpha_{12}-6)} = \left[\frac{\epsilon_{11}\alpha_{11}\sigma_{11}^6}{(\alpha_{11}-6)}\frac{\epsilon_{22}\alpha_{22}\sigma_{22}^6}{(\alpha_{22}-6)}\right]^{\frac{1}{2}}$$

Other rules for that have been proposed for the Exp-6 potential are the Mason-Rice rules and the Srivastava and Srivastava rules (1956).

References

- Beauchamp, KA; Lin, YS; Das, R; Pande, VS (10 April 2012). "Are Protein Force Fields Getting Better? A Systematic Benchmark on 524 Diverse NMR Measurements.". Journal of chemical theory and computation

- Gillan, M. J. (1990). "The path-integral simulation of quantum systems, Section 2.4". In C. R. A. Catlow, S. C. Parker and M. P. Allen. Computer Modelling of Fluids Polymers and Solids. NATO ASI Series C. 293. pp. 155–188. ISBN 978-0-7923-0549-1

- Alder, B. J.; Wainwright, T. E. (1959). "Studies in Molecular Dynamics. I. General Method". J. Chem. Phys. 31 (2): 459. Bibcode:1959JChPh..31..459A. doi:10.1063/1.1730376

- "The Nobel Prize in Chemistry 2013" (PDF) (Press release). Royal Swedish Academy of Sciences. October 9, 2013. Retrieved October 9, 2013

- Cao, J.; Voth, G. A. (1996). "Semiclassical approximations to quantum dynamical time correlation functions". The Journal of Chemical Physics. 104: 273. Bibcode:1996JChPh.104..273C. doi:10.1063/1.470898

- Binder, K. & Heermann, D.W. (2002). Monte Carlo Simulation in Statistical Physics. An Introduction (4th edition). Springer. ISBN 3-540-43221-3

- Chakrabarty, A; T Cagin (2010). "Coarse grain modeling of polyimide copolymers". Polymer. 51 (12): 2786–2794. doi:10.1016/j.polymer.2010.03.060

- Gervasio, F.; Laio, A.; Parrinello, M. (2005). "Flexible docking in solution using metadynamics". Journal of the American Chemical Society. 127 (8): 2600–2607. PMID 15725015. doi:10.1021/ja0445950

- Sommerfeld, Arnold (1952). Lectures on Theoretical Physics, Vol. I: Mechanics. New York: Academic Press. ISBN 0-12-654670-3

- Bazant, M. Z.; Kaxiras, E.; Justo, J. F. (1997). "Environment-dependent interatomic potential for bulk silicon". Phys. Rev. B. 56: 8542. doi:10.1103/PhysRevB.56.8542

- Mazur, AK (1999). "Symplectic integration of closed chain rigid body dynamics with internal coordinate equations of motion". Journal of Chemical Physics. 111 (4): 1407–1414. Bibcode:-1999JChPh.111.1407M. doi:10.1063/1.479399

- Schaefer M, van Vlijmen HW, Karplus M (1998). "Electrostatic contributions to molecular free energies in solution". Advances in Protein Chemistry. Advances in Protein Chemistry. 51: 1–57. ISBN 978-0-12-034251-8. PMID 9615168. doi:10.1016/S0065-3233(08)60650-6

- Berne, B. J.; Thirumalai, D. (1986). "On the Simulation of Quantum Systems: Path Integral Methods". Annual Review of Physical Chemistry. 37: 401. Bibcode:1986ARPC...37..401B. doi:10.1146/annurev.pc.37.100186.002153

- Chang, Kenneth (October 9, 2013). "3 Researchers Win Nobel Prize in Chemistry". New York Times. Retrieved October 9, 2013

- Walter A, Gutknecht J (1986). "Permeability of small nonelectrolytes through lipid bilayer membranes". The Journal of Membrane Biology. 90 (3): 207–17. PMID 3735402. doi:10.1007/BF01870127

Force Field: An Integrated Study

Force fields in molecular modeling refer to the functional form and parameter sets that are established in order to compute the potential energy of a system of atoms. Classical force fields include Assisted Model Building and Energy Refinement (AMBER), Chemistry at HARvard Molecular Mechanics (CHARMM), etc. Computational chemistry is best understood in confluence with the major topics listed in the following chapter.

Force Field (Chemistry)

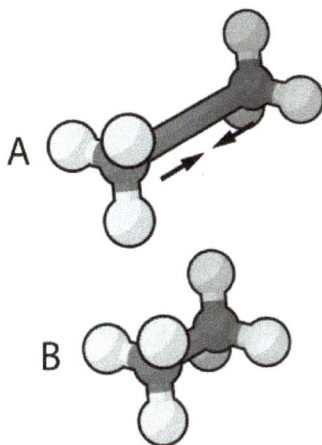

A force field is used to minimize the bond stretching energy of this ethane molecule.

In the context of molecular modeling, a force field (a special case of energy functions or interatomic potentials; but different than force field in classical physics) refers to the functional form and parameter sets used to calculate the potential energy of a system of atoms or coarse-grained particles in molecular mechanics and molecular dynamics simulations. The parameters of the energy functions may be derived from experiments in physics or chemistry, calculations in quantum mechanics, or both.

All-atom force fields provide parameters for every type of atom in a system, including hydrogen, while *united-atom* interatomic potentials treat the hydrogen and carbon atoms in each methyl group (terminal methyl) and each methylene bridge as one interaction center. *Coarse-grained* potentials, which are often used in long-time simulations of macromolecules such as proteins, nucleic acids, and multi-component complexes, provide even cruder representations for higher computing efficiency.

Functional Form

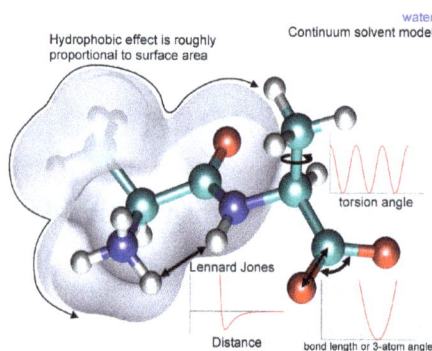

Molecular mechanics potential energy function with continuum solvent

The basic functional form of potential energy in molecular mechanics includes bonded terms for interactions of atoms that are linked by covalent bonds, and nonbonded (also termed *noncovalent*) terms that describe the long-range electrostatic and van der Waals forces. The specific decomposition of the terms depends on the force field, but a general form for the total energy in an additive force field can be written as $E_{total} = E_{bonded} + E_{nonbonded}$ where the components of the covalent and noncovalent contributions are given by the following summations:

$$E_{bonded} = E_{bond} + E_{angle} + E_{dihedral}$$

$$E_{nonbonded} = E_{electrostatic} + E_{van\ der\ Waals}$$

The bond and angle terms are usually modeled by quadratic energy functions that do not allow bond breaking. A more realistic description of a covalent bond at higher stretching is provided by the more expensive Morse potential. The functional form for dihedral energy is highly variable. Additional, "improper torsional" terms may be added to enforce the planarity of aromatic rings and other conjugated systems, and "cross-terms" that describe coupling of different internal variables, such as angles and bond lengths. Some force fields also include explicit terms for hydrogen bonds.

The nonbonded terms are most computationally intensive. A popular choice is to limit interactions to pairwise energies. The van der Waals term is usually computed with a Lennard-Jones potential and the electrostatic term with Coulomb's law, although both can be buffered or scaled by a constant factor to account for electronic polarizability and produce better agreement with experimental observations.

Parameterization

In addition to the functional form of the potentials, force fields define a set of parameters for different types of atoms, chemical bonds, dihedral angles and so on. The parameter sets are usually empirical. A force field would include distinct parameters for an oxygen atom in a carbonyl functional group and in a hydroxyl group. The typical parameter set includes values for atomic mass, van der Waals radius, and partial charge for individual atoms, and equilibrium values of bond lengths, bond angles, and dihedral angles for pairs, triplets, and quadruplets of bonded atoms, and values corresponding to the effective spring constant for each potential. Most current force fields parameters use a *fixed-charge* model by which each atom is assigned one value for the atomic charge that is not affected by the local electrostatic environment; proposed developments in next-generation force fields incorporate models for polarizability, in which a particle's charge is influenced by electrostatic interactions with its neighbors. For example, polarizability can be approximated by the introduction of induced dipoles; it can also be represented by Drude particles, massless, charge-carrying virtual sites attached by a springlike harmonic oscillator potential to each polarizable atom. The introduction of polarizability into force fields in common use has been inhibited by the high computational expense associated with calculating the local electrostatic field.

Although many molecular simulations involve biological macromolecules such as proteins, DNA, and RNA, the parameters for given atom types are generally derived from observations on small organic molecules that are more tractable for experimental studies and quantum calculations. Different force field parameters can be derived from dissimilar types of experimental data, such as enthalpy of vaporization (OPLS), enthalpy of sublimation, dipole moments, or various spectroscopic parameters.

Parameter sets and functional forms are defined by interatomic potentials developers to be self-consistent. Because the functional forms of the potential terms vary extensively between even closely related interatomic potentials (or successive versions of the same interatomic potential), the parameters from one interatomic potential function should clearly never be used together with another interatomic potential function.

Deficiencies

All interatomic potentials are based on many approximations and derived from different types of experimental data. Thus, they are termed *empirical*. Some existing energy functions do not account for electronic polarization of the environment, an effect that can significantly reduce electrostatic interactions of partial atomic charges. This problem was addressed by developing *polarizable force fields* or using macroscopic dielectric constant. However, application of one value of dielectric constant is questionable in the highly heterogeneous environments of proteins or biological membranes, and the nature of the dielectric depends on the model used.

All types of van der Waals forces are also strongly environment-dependent, because these forces originate from interactions of induced and "instantaneous" dipoles. The original Fritz London theory of these forces can only be applied in vacuum. A more general theory of van der Waals forces in condensed media was developed by A. D. McLachlan in 1963 (this theory includes the original London's approach as a special case).

The McLachlan theory predicts that van der Waals attractions in media are weaker than in vacuum and follow the *like dissolves like* rule, which means that different types of atoms interact more weakly than identical types of atoms. This is in contrast to *combinatorial rules* or Slater-Kirkwood equation applied for development of the classical force fields. The *combinatorial rules* state that interaction energy of two dissimilar atoms (e.g., C...N) is an average of the interaction energies of corresponding identical atom pairs (i.e., C...C and N...N). According to McLachlan theory, the interactions of particles in a media can even be fully repulsive, as observed for liquid helium. The conclusions of McLachlan theory are supported by direct measurements of attraction forces between different materials (Hamaker constant), as explained by Jacob Israelachvili in his book *Intermolecular and surface forces*. It was concluded that *"the interaction between hydrocarbons across water is about 10% of that across vacuum"*. Such effects are unaccounted in standard molecular mechanics.

Another round of criticism came from practical applications, such as protein structure refinement. It was noted that *Critical Assessment of protein Structure Prediction* (CASP) participants did not try to refine their models to avoid *"a central embarrassment of molecular mechanics, namely that energy minimization or molecular dynamics generally leads to a model that is less like the experimental structure"*. The force fields have been applied successfully for protein structure refinement in different X-ray crystallography and NMR spectroscopy applications, especially using program XPLOR. However, such refinement is driven mainly by a set of experimental constraints, whereas the interatomic potentials serve merely to remove interatomic hindrances. The results of calculations are practically the same with rigid sphere potentials implemented in program DYANA (calculations from NMR data), or with programs for crystallographic refinement that do not use any energy functions. The deficiencies of the interatomic potentials remain a major bottleneck in homology modeling of proteins. Such situation gave rise to development of alternative empirical scoring functions specifically for ligand docking, protein folding, homology model refinement, computational protein design, and modeling of proteins in membranes.

There is also an opinion that molecular mechanics may operate with energy which is irrelevant to protein folding or ligand binding. The parameters of typical force fields reproduce enthalpy of sublimation, i.e., energy of evaporation of molecular crystals. However, it was recognized that protein folding and ligand binding are thermodynamically very similar to crystallization, or liquid-solid transitions, because all these processes represent *freezing* of mobile molecules in condensed media. Thus, free energy changes during protein folding or ligand binding are expected to represent a combination of an energy similar to heat of fusion (energy absorbed during melting of molecular crystals), a conformational entropy contribution, and solvation free energy. The heat of fusion is significantly smaller than enthalpy of sublimation. Hence, the potentials describing protein folding or ligand binding must be weaker than potentials in molecular mechanics. Indeed, the energies of H-bonds in proteins are ~ -1.5 kcal/mol when estimated from protein engineering or alpha helix to coil transition data, but the same energies estimated from sublimation enthalpy of molecular crystals were -4 to -6 kcal/mol. The depths of modified Lennard-Jones potentials derived from protein engineering data were also smaller than in typical potential parameters and followed the *like dissolves like* rule, as predicted by McLachlan theory.

Future Perspectives

The use of interatomic potentials in chemistry was first introduced in 1949, apparently independently by Hill and by Westheimer, applied mainly to organic chemistry to estimate properties such as strain energies among others. The functional form of the interatomic potential, applied to biological systems, was established by Lifson in the 1960s. For over a half century, interatomic potentials have served us well, providing useful insights into and interpretation of biomolecular structure and function. Undoubtedly, it will continue to be widely used, thanks to its computational efficiency, while its reliability will continue to be improved. Yet, there are many well-known deficiencies as noted. Further, the number of energy terms used in a given interatomic potential cannot be uniquely determined and a highly redundant number of degrees of freedom are typically used. Consequently, the "parameters" in different interatomic potentials can be vastly different. Of course, the emphasis to incorporate polarization into the standard pair-wise potentials can be very useful; however, there is no unique way of treating polarization in molecular mechanics because it is of quantum mechanical origin. Furthermore, often we are more interested in the properties derived from the dynamic dependence of the interatomic potential itself on molecular fluctuations.

One possibility is that the future development of interatomic potential ought to move beyond the current molecular mechanics approach, by using quantum mechanics explicitly to construct the interatomic potential. A number of the *polarizable interatomic potentials* listed, such as density fitting and bond-polarization, already included some of the key ingredients towards this goal. The explicit polarization (X-Pol) method appears to have established the fundamental theoretical framework for a quantal force field; the next step is to develop the necessary parameters to achieve more accurate results than classical mechanics can offer.

Popular Force Fields

Different force fields are designed for different purposes. All are implemented in various computer software.

MM2 was developed by Norman Allinger mainly for conformational analysis of hydrocarbons and other small organic molecules. It is designed to reproduce the equilibrium covalent geometry of molecules as precisely as possible. It implements a large set of parameters that is continuously refined and updated for many different classes of organic compounds (MM3 and MM4).

CFF was developed by Arieh Warshel, Lifson and coworkers as a general method for unifying studies of energies, structures and vibration of general molecules and molecular crystals. The CFF program, developed by Levitt and Warshel, is based on the Cartesian representation of all the atoms, and it served as the basis for many subsequent simulation programs.

ECEPP was developed specifically for modeling of peptides and proteins. It uses fixed geometries of amino acid residues to simplify the potential energy surface. Thus, the energy minimization is conducted in the space of protein torsion angles. Both MM2 and ECEPP include potentials for H-bonds and torsion potentials for describing rotations around single bonds. ECEPP/3 was implemented (with some modifications) in Internal Coordinate Mechanics and FANTOM.

AMBER, CHARMM, and GROMOS have been developed mainly for molecular dynamics of macromolecules, although they are also commonly used for energy minimizing. Thus, the coordinates of all atoms are considered as free variables.

Classical Force Fields

- Assisted Model Building and Energy Refinement (AMBER) – widely used for proteins and DNA.

- Chemistry at HARvard Molecular Mechanics (CHARMM) – originally developed at Harvard, widely used for both small molecules and macromolecules

- CVFF – also used broadly for small molecules and macromolecules.

- COSMOS-NMR – hybrid QM/MM force field adapted to a variety of inorganic compounds, organic compounds and biological macromolecules, including semi-empirical calculation of atomic charges and NMR properties. COSMOS-NMR is optimized for NMR based structure elucidation and implemented in COSMOS molecular modelling package.

- GROningen MOlecular Simulation (GROMOS) – a force field that comes as part of the GROMOS software, a general-purpose molecular dynamics computer simulation package for the study of biomolecular systems. GROMOS force field A-version has been developed for application to aqueous or apolar solutions of proteins, nucleotides, and sugars. A B-version to simulate gas phase isolated molecules is also available.

- Optimized Potential for Liquid Simulations (OPLS, variants include OPLS-AA, OPLS-UA, OPLS-2001, OPLS-2005) – developed by William L. Jorgensen at the Yale University Department of Chemistry.

- ECEPP – first force field for polypeptide molecules - developed by F.A. Momany, H.A. Scheraga and colleagues.

- QCFF/PI – A general force fields for conjugated molecules.

- Universal Force Field (UFF) – A general force field with parameters for the full periodic table up to and including the actinoids, developed at Colorado State University.

- Consistent Force Field (CFF) – a family of forcefields adapted to a broad variety of organic compounds, includes force fields for polymers, metals, etc.

- Condensed-phase Optimized Molecular Potentials for Atomistic Simulation Studies (COMPASS) – developed by H. Sun at Molecular Simulations Inc., parameterized for a variety of molecules in the condensed phase, now available through Accelrys.

- Merck Molecular Force Field (MMFF) – developed at Merck, for a broad range of molecules.

- MM2, MM3, MM4 – developed by Norman Allinger, parametrized for a broad range of molecules.

- QVBMM - developed by Vernon G. S. Box, parameterized for all biomolecules and a broad range of organic molecules, and implemented in StruMM3D (STR-3DI32).

- Transferable Potentials for Phase Equilibria (TraPPE) – a family of molecular mechanics force fields developed by the Siepmann group at the University of Minnesota for molecular simulations of complex chemical systems.

Polarizable Force Fields

- X-Pol: the Explicit Polarization Theory – a fragment-based electronic structure method introduced by Jiali Gao at the University of Minnesota, which can be used at any level of theory—ab initio Hartree–Fock method (HF), semiempirical molecular orbital theory, correlated wave function theory, or Kohn-Sham (KS) density functional theory (DFT). It can perform over 3,200 steps (3.2 ps) of MD simulations of a fully solvated protein in water with periodic boundary conditions, consisting of about 15,000 atoms and 30,000 basis functions on one processor in 24 hours in 2008, with a full quantum mechanical representation of the whole system. Note that the first MD simulation of a protein by McCammon, Gelin, and Karplus in 1977 lasted 8.8 ps using a united-atom force field without solvent.

- CFF/ind and ENZYMIX – The first polarizable force field which has subsequently been used in many applications to biological systems.

- DRF90 developed by P. Th. van Duijnen and coworkers.

- PIPF – The polarizable intermolecular potential for fluids is an induced point-dipole force field for organic liquids and biopolymers. The molecular polarization is based on Thole's interacting dipole (TID) model and was developed by Jiali Gao at the University of Minnesota.

- Polarizable Force Field (PFF) – developed by Richard A. Friesner and coworkers.

- SP-basis Chemical Potential Equalization (CPE) – approach developed by R. Chelli and P. Procacci.

- CHARMM – polarizable force field developed by S. Patel (University of Delaware) and C. L. Brooks III (University of Michigan).

- AMBER – polarizable force field developed by Jim Caldwell and coworkers.

- CHARMM – polarizable force field based on the classical Drude oscillator developed by A. MacKerell (University of Maryland, Baltimore) and B. Roux (University of Chicago).

- Sum of Interactions Between Fragments Ab initio computed (SIBFA) – force field for small molecules and flexible proteins, developed by Nohad Gresh (Paris V, René Descartes University) and Jean-Philip Piquemal (Paris VI, Pierre & Marie Curie University). SIBFA is a molecular mechanics procedure formulated and calibrated on the basis of ab initio supermolecule computations. Its purpose is to enable the simultaneous and reliable computations of both intermolecular and conformational energies governing the binding specificities of biologically and pharmacologically relevant molecules. This procedure enables an accurate treatment of transition metals. The inclusion of a ligand field contribution allows computations on "open-shell" metalloproteins.

- Atomic Multipole Optimized Energetics for Biomolecular Applications (AMOEBA) – force field developed by Pengyu Ren (University of Texas at Austin) and Jay W. Ponder (Washington University).

- ORIENT – procedure developed by Anthony J. Stone (Cambridge University) and coworkers.

- Non-Empirical Molecular Orbital (NEMO) – procedure developed by Gunnar Karlström and coworkers at Lund University (Sweden).

- Gaussian Electrostatic Model (GEM) – a polarizable force field based on Density Fitting developed by Thomas A. Darden and G. Andrés Cisneros at NIEHS; and Jean-Philip Piquemal at Paris VI University.

- Polarizable procedure based on the Kim-Gordon approach developed by Jürg Hutter and coworkers (University of Zürich).

- Computer Simulation of Molecular Structure (COSMOS-NMR) – developed by Ulrich Sternberg and coworkers. Hybrid QM/MM force field enables explicit quantum-mechanical calculation of electrostatic properties using localized bond orbitals with fast BPT formalism. Atomic charge fluctuation is possible in each molecular dynamics step.

Reactive Force Fields

- ReaxFF – reactive force field (interatomic potential) developed by Adri van Duin, William Goddard and coworkers. It is fast, transferable and is the computational method of choice for atomistic-scale dynamical simulations of chemical reactions. Parallelized ReaxFF allows reactive simulations on >>1,000,000 atoms.

- Empirical valence bond (EVB) – this reactive force field, introduced by Warshel and coworkers, is probably the most reliable and physically consistent way to use force fields in modeling chemical reactions in different environments. The EVB facilitates calculating activation free energies in condensed phases and in enzymes.

- RWFF – reactive force field for water developed by Detlef W. M. Hofmann, Liudmila N. Kuleshova, and Bruno D'Aguanno. It is very fast, reproduces the experimental data of neutron scattering accurately, and allows simuling bond formation-breaking of water and acids.

Coarse-grained Force Fields

- Virtual atom molecular mechanics (VAMM) – a coarse-grained force field developed by Korkut and Hendrickson for molecular mechanics calculations such as large scale conformational transitions based on the virtual interactions of C-alpha atoms. It is a knowledge based force field and formulated to capture features dependent on secondary structure and on residue-specific contact information in proteins.

- MARTINI – a coarse-grained potential developed by Marrink and coworkers at the University of Groningen, initially developed for molecular dynamics simulations of lipids, later extended to various other molecules. The force field applies a mapping of four heavy atoms to one CG interaction site and is parameterized with the aim of reproducing thermodynamic properties.

- SIRAH – a coarse-grained force field developed by Pantano and coworkers of the Biomolecular Simulations Group, Institut Pasteur of Montevideo, Uruguay; developed for molecular dynamics of water, DNA and proteins. Free available for AMBER and GROMACS packages.

Water Models

The set of parameters used to model water or aqueous solutions (basically a force field for water) is called a water model. Water has attracted a great deal of attention due to its unusual properties and its importance as a solvent. Many water models have been proposed; some examples are TIP3P, TIP4P, SPC, flexible simple point charge water model (flexible SPC), and ST2.

Post-translational Modifications and Unnatural Amino Acids

- Forcefield_PTM – An AMBER-based forcefield and webtool for modeling common post-translational modifications of amino acids in proteins developed by Chris Floudas and coworkers. It uses the ff03 charge model and has several side-chain torsion corrections parameterized to match the quantum chemical rotational surface.

- Forcefield_NCAA - An AMBER-based forcefield and webtool for modeling common non-natural amino acids in proteins in condensed-phase simulations using the ff03 charge model. The charges have been reported to be correlated with hydration free energies of corresponding side-chain analogs.

Other

- VALBOND - a function for angle bending that is based on valence bond theory and works for large angular distortions, hypervalent molecules, and transition metal complexes. It can be incorporated into other force fields such as CHARMM and UFF.

MARTINI

Martini is a coarse-grained (CG) force field developed by Marrink and coworkers at the University of Groningen, initially developed in 2004 for molecular dynamics simulation of lipids, later (2007) extended to various other molecules. The force field applies a mapping of four heavy atoms to one CG interaction site and is parametrized with the aim of reproducing thermodynamic properties.

Philosophy

For the Martini force field 4 bead categories have been defined: Q (charged), P (polar), N (nonpolar), and C (apolar). These bead types are in turn split in 4 or 5 different levels, giving a total of 20 beadtypes. For the interactions between the beads, 10 different interaction levels are defined (O-IX). The beads can be used at normal size (4:1 mapping) or S-size (small, 3:1 mapping). The latter is mainly used in ring structures. Bonded interactions (bonds, angles, dihedrals, and impropers) are derived from atomistic simulations of crystal structures.

Use

The Martini force field has become one of the most used coarse grained force fields in the field of molecular dynamics simulations for biomolecules. The original 2004 and 2007 papers have been cited 654 and 608 times, respectively. The force field has been implemented in three major simulation codes: GROningen MAchine for Chemical Simulations (GROMACS), GROningen MOlecular Simulation (GROMOS), and Nanoscale Molecular Dynamics (NAMD). Notable successes are simulations of the clustering behavior of syntaxin-1A, the simulations of the opening of mechanosensitive channels (MscL) and the simulation of the domain partitioning of membrane peptides.

Parameter Sets

Lipids

The initial papers contained parameters for water, simple alkanes, organic solvents, surfactants, a wide range of lipids and cholesterol. They semiquantitatively reproduce the phase behavior of bilayers with other bilayer properties, and more complex bilayer behavior.

Proteins

Compatible parameters for proteins were introduced by Monticelli *et al.*. Secondary structure elements, like alpha helixes and beta sheets (β-sheets), are constrained. Martini proteins are often simulated in combination with an elastic network, such as Elnedyn, to maintain the overall structure.

CHARMM

Chemistry at Harvard Macromolecular Mechanics (CHARMM) is the name of a widely used set of force fields for molecular dynamics, and the name for the molecular dynamics simulation and analysis computer software package associated with them. The CHARMM Development Project involves a worldwide network of developers working with Martin Karplus and his group at Harvard to develop and maintain the CHARMM program. Licenses for this software are available, for a fee, to people and groups working in academia.

Force Fields

The CHARMM force fields for proteins include: united-atom (sometimes termed *extended atom*) CHARMM19, all-atom CHARMM22 and its dihedral potential corrected variant CHARMM22/CMAP. In the CHARMM22 protein force field, the atomic partial charges were derived from quantum chemical calculations of the interactions between model compounds and water. Furthermore, CHARMM22 is parametrized for the TIP3P explicit water model. Nevertheless, it is often used with implicit solvents. In 2006, a special version of CHARMM22/CMAP was reparametrized for consistent use with implicit solvent GBSW.

For DNA, RNA, and lipids, CHARMM27 is used. Some force fields may be combined, for example CHARMM22 and CHARMM27 for the simulation of protein-DNA binding. Also, parameters for NAD+, sugars, fluorinated compounds, etc., may be downloaded. These force field version numbers refer to the CHARMM version where they first appeared, but may of course be used with subsequent versions of the CHARMM executable program. Likewise, these force fields may be used within other molecular dynamics programs that support them.

In 2009, a general force field for drug-like molecules (CGenFF) was introduced. It "covers a wide range of chemical groups present in biomolecules and drug-like molecules, including a large number of heterocyclic scaffolds". The general force field is designed to cover any combination of chemical groups. This inevitably comes with a decrease in accuracy for representing any particular subclass of molecules. Users are repeatedly warned in Mackerell's website not to use the CGenFF parameters for molecules for which specialized force fields already exist (as mentioned above for proteins, nucleic acids, etc.).

CHARMM also includes polarizable force fields using two approaches. One is based on the fluctuating charge (FQ) model, also termed Charge Equilibration (CHEQ). The other is based on the Drude shell or dispersion oscillator model.

Parameters for all of these force fields may be downloaded from the Mackerell website for free.

Molecular Dynamics Program

The CHARMM program allows generating and analysing a wide range of molecular simulations. The most basic kinds of simulation are minimizing a given structure and production runs of a molecular dynamics trajectory.

More advanced features include free energy perturbation (FEP), quasi-harmonic entropy estimation, correlation analysis and combined quantum, and molecular mechanics (QM/MM) methods.

CHARMM is one of the oldest programs for molecular dynamics. It has accumulated many features, some of which are duplicated under several keywords with slight variants. This is an inevitable result of the many outlooks and groups working on CHARMM worldwide. The changelog file, and CHARMM's source code, are good places to look for the names and affiliations of the main developers. The involvement and coordination by Charles L. Brooks III's group at the University of Michigan is salient.

Software History

Around 1969, there was considerable interest in developing potential energy functions for small molecules. CHARMM originated at Martin Karplus's group at Harvard. Karplus and his then graduate student Bruce Gelin decided the time was ripe to develop a program that would make it possible to take a given amino acid sequence and a set of coordinates (e.g., from the X-ray structure) and to use this information to calculate the energy of the system as a function of the atomic positions. Karplus has acknowledged the importance of major inputs in the development of the (at the time nameless) program, including:

- Schneior Lifson's group at the Weizmann Institute, especially from Arieh Warshel who went to Harvard and brought his consistent force field (CFF) program with him

- Harold Scheraga's group at Cornell University

- Awareness of Michael Levitt's pioneering energy calculations for proteins

In the 1980s, finally a paper appeared and CHARMM made its public début. Gelin's program had by then been considerably restructured. For the publication, Bob Bruccoleri came up with the name HARMM (HARvard Macromolecular Mechanics), but it seemed inappropriate. So they added a C for Chemistry. Karplus said: "*I sometimes wonder if Bruccoleri's original suggestion would have served as a useful warning to inexperienced scientists working with the program.*" CHARMM has continued to grow and the latest release of the executable program was made in August 2009 as CHARMM35b3.

Running Charmm Under Unix-Linux

The general syntax for using the program is:

charmm -i filename.inp -o filename.out

- charmm – The name of the program (or script which runs the program) on the computer system being used.

- filename.inp – A text file which contains the CHARMM commands. It starts by loading the molecular topologies (top) and force field (par). Then one loads the molecular structures' Cartesian coordinates (e.g. from PDB files). One can then modify the molecules (adding hydrogens, changing secondary structure). The calculation section can include energy minimization, dynamics production, and analysis tools such as motion and energy correlations.

- filename.out – The log file for the CHARMM run, containing echoed commands, and various amounts of command output. The output print level may be increased or decreased in general, and procedures such as minimization and dynamics have printout frequency specifications. The values for temperature, energy pressure, etc. are output at that frequency.

Volunteer Computing

Docking@Home, hosted by University of Delaware, one of the projects which use an open source platform for the distributed computing, BOINC, used CHARMM to analyze the atomic details of protein-ligand interactions in terms of molecular dynamics (MD) simulations and minimizations.

World Community Grid, sponsored by IBM, ran a project named The Clean Energy Project which also used CHARMM in its first phase which has completed.

Gromos

GROMOS is the name of a force field for molecular dynamics simulation, and a related computer software package. Both are developed at the University of Groningen, and at the Computer-Aided Chemistry Group at the Laboratory for Physical Chemistry at the Swiss Federal Institute of Technology (ETH Zurich). At Groningen, Herman Berendsen was involved in its development.

The united atom force field was optimized with respect to the condensed phase properties of alkanes.

Versions

GROMOS87

Aliphatic and aromatic hydrogen atoms were included implicitly by representing the carbon atom and attached hydrogen atoms as one group centered on the carbon atom, a united atom force field. The van der Waals force parameters were derived from calculations of the crystal structures of hydrocarbons, and on amino acids using short (0.8 nm) nonbonded cutoff radii.

GROMOS96

In 1996, a substantial rewrite of the software package was released. The force field was also improved, e.g., in the following way: aliphatic CH_n groups were represented as united atoms with van der Waals interactions reparametrized on the basis of a series of molecular dynamics simulations of model liquid alkanes using long (1.4 nm) nonbonded cutoff radii. This version is continually being refined and several different parameter sets are available. GROMOS96 includes studies of molecular dynamics, stochastic dynamics, and energy minimization. The energy component was also part of the prior GROMOS, named GROMOS87. GROMOS96 was planned and conceived during a time of 20 months. The package is made of 40 different programs, each with a different essential function. An example of two important programs within the GROMOS96 are PROGMT, in charge of constructing molecular topology and also PROPMT, changing the classical molecular topology into the path-integral molecular topology.

GROMOS05

An updated version of the software package was introduced in 2005.

GROMOS11

The current GROMOS release is dated in May 2011.

Parameter Sets

Some of the force field parameter sets that are based on the GROMOS force field. The A-version applies to aqueous or apolar solutions of proteins, nucleotides, and sugars. The B-version applies to isolated molecules (gas phase).

54

- 54A7 - 53A6 taken and adjusted torsional angle terms to better reproduce helical propensities, altered N–H, C=O repulsion, new CH_3 charge group, parameterisation of Na^+ and Cl^- to improve free energy of hydration and new improper dihedrals.

- 54B7 - 53B6 *in vacuo* taken and changed in same manner as 53A6 to 54A7.

53

- 53A5 - optimised by first fitting to reproduce the thermodynamic properties of pure liquids of a range of small polar molecules and the solvation free enthalpies of amino acid analogs in cyclohexane, is an expansion and renumbering of 45A3.

- 53A6 - 53A5 taken and adjusted partial charges to reproduce hydration free enthalpies in water, recommended for simulations of biomolecules in explicit water.

45

- 45A3 - suitable to apply to lipid aggregates such as membranes and micelles, for mixed systems of aliphatics with or without water, for polymers, and other apolar systems that may interact with different biomolecules.

- 45A4 - 45A3 reparameterised to improve DNA representation.

43

- 43A1

- 43A2

ReaxFF

ReaxFF (for "reactive force field") is a bond order based force field developed by Adri van Duin, William A. Goddard, III, and co- workers at the California Institute of Technology for use, e.g., in molecular dynamics simulations. Whereas traditional force fields are unable to model chemical reactions because of the requirement of breaking and forming bonds (a force field's functional form depends on having all bonds defined explicitly), ReaxFF eschews explicit bonds in favor of bond orders, which allows for continuous bond formation/breaking. ReaxFF aims to be as general as possible and has been parameterized and tested for hydrocarbon reactions, alkoxysilane gelation, transition-metal-catalyzed nanotube formation, and high-energy materials.

Recently, ReaxFF has been developed to study oxygen interactions with realistic silica surfaces. This version of ReaxFF is based on highly accurate and benchmarking density functional studies. Highly accurate density functional results are achieved by employing Minnesota Functionals.

AMBER

Assisted Model Building with Energy Refinement (AMBER) is a family of force fields for molecular dynamics of biomolecules originally developed by Peter Kollman's group at the University of California, San Francisco. AMBER is also the name for the molecular dynamics software package that simulates these force fields. It is maintained by an active collaboration between David Case at Rutgers University, Tom Cheatham at the University of Utah, Tom Darden at NIEHS, Ken Merz at Michigan State University, Carlos Simmerling at Stony Brook University, Ray Luo at UC Irvine, and Junmei Wang at Encysive Pharmaceuticals.

Force Field

The term *AMBER force field* generally refers to the functional form used by the family of AMBER force fields. This form includes several parameters; each member of the family of AMBER force fields provides values for these parameters and has its own name.

Functional Form

The functional form of the AMBER force field is

$$V(r^N) = \sum_{bonds} k_b (l - l_0)^2 + \sum_{angles} k_a (\theta - \theta_0)^2$$

$$+ \sum_{torsions} \sum_n \frac{1}{2} V_n [1 + \cos(n\omega - \gamma)] + \sum_{j=1}^{N-1} \sum_{i=j+1}^{N} f_{ij} \left\{ \epsilon_{ij} \left[\left(\frac{r_{0ij}}{r_{ij}} \right)^{12} - 2 \left(\frac{r_{0ij}}{r_{ij}} \right)^6 \right] + \frac{q_i q_j}{4\pi\epsilon_0 r_{ij}} \right\}$$

Despite the term *force field*, this equation defines the potential energy of the system; the force is the derivative of this potential relative to position.

The meanings of right hand side terms are:

- First term (summing over bonds): represents the energy between covalently bonded atoms. This harmonic (ideal spring) force is a good approximation near the equilibrium bond length, but becomes increasingly poor as atoms separate.

- Second term (summing over angles): represents the energy due to the geometry of electron orbitals involved in covalent bonding.

- Third term (summing over torsions): represents the energy for twisting a bond due to bond order (e.g., double bonds) and neighboring bonds or lone pairs of electrons. One bond may have more than one of these terms, such that the total torsional energy is expressed as a Fourier series.

- Fourth term (double summation over i and j): represents the non-bonded energy between all atom pairs, which can be decomposed into van der Waals (first term of summation) and electrostatic (second term of summation) energies.

The form of the van der Waals energy is calculated using the equilibrium distance (r_{0ij}) and well depth (ϵ). The factor of 2 ensures that the equilibrium distance is r_{0ij}. The energy is sometimes reformulated in terms of σ, where $r_{0ij} = 2^{1/6}(\sigma),$, as used e.g. in the implementation of the softcore potentials.

The form of the electrostatic energy used here assumes that the charges due to the protons and electrons in an atom can be represented by a single point charge (or in the case of parameter sets that employ lone pairs, a small number of point charges).

Parameter Sets

To use the AMBER force field, it is necessary to have values for the parameters of the force field (e.g. force constants, equilibrium bond lengths and angles, charges). A fairly large number of these parameter sets exist, and are described in detail in the AMBER software user manual. Each parameter set has a name, and provides parameters for certain types of molecules.

- Peptide, protein, and nucleic acid parameters are provided by parameter sets with names starting with "ff" and containing a two digit year number, for instance "ff99".

- *General AMBER force field* (GAFF) provides parameters for small organic molecules to facilitate simulations of drugs and small molecule ligands in conjunction with biomolecules.

- The GLYCAM force fields have been developed by Rob Woods for simulating carbohydrates.

Software

The AMBER software suite provides a set of programs to apply the AMBER forcefields to simulations of biomolecules. It is written in the programming languages Fortran 90 and C, with support for most major Unix-like operating systems and compilers. Development is conducted by a loose association of mostly academic labs. New versions are released usually in the spring of even numbered years; AMBER 10 was released in April 2008. The software is available under a site license agreement, which includes full source, currently priced at US$500 for non-commercial and US$20,000 for commercial organizations.

Programs

- *LEaP* prepares input files for the simulation programs.

- *Antechamber* automates the process of parameterizing small organic molecules using GAFF.

- *Simulated Annealing with NMR-Derived Energy Restraints* (SANDER) is the central simulation program and provides facilities for energy minimizing and molecular dynamics with a wide variety of options.

- *pmemd* is a somewhat more feature-limited reimplementation of SANDER by Bob Duke. It was designed for parallel computing, and performs significantly better than SANDER when running on more than 8–16 processors.

  o *pmemd.cuda* runs simulations on machines with graphics processing units (GPUs).

  o *pmemd.amoeba* handles the extra parameters in the polarizable AMOEBA force field.

- *nmode* calculates normal modes.

- *ptraj* numerically analyzes simulation results. AMBER includes no visualizing abilities, which is commonly performed with Visual Molecular Dynamics (VMD). Ptraj is now unsupported as of AmberTools 13.

- *cpptraj* is a rewritten version of ptraj made in C++ to give faster analysis of simulation results. Several actions have been made parallelizable with OpenMP and MPI.

- *MM-PBSA* allows implicit solvent calculations on snap shots from molecular dynamics simulations.

- *NAB* is a built-in nucleic acid building environment made to aid in the process of manipulating proteins and nucleic acids where an atomic level of description will aid computing.

Merck Molecular Force Field

Merck Molecular Force Field (MMFF) is a family of chemistry force fields developed by Merck Research Laboratories. They are based on the MM3 force field. MMFF is not optimized for one use, such as simulating proteins or small molecules, but tries to perform well for a wide range of organic chemistry calculations. The parameters in the force field have been derived from computational data.

The first published force field in the family is MMFF94. A set of molecular structures and the corresponding output of Halgren's MMFF94 implementation is provided at the Computational Chemistry List for validating other MMFF implementations.

TraPPE Force Field

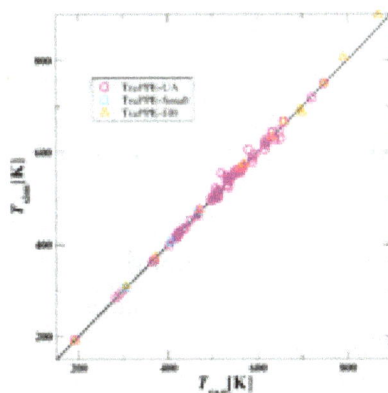

Graph of TraPPE force field accuracy relative to critical temperatures

Transferable Potentials for Phase Equilibria (TraPPE) is a family of molecular mechanics force fields developed mainly by the research group of J. Ilja Siepmann at the University of Minnesota. The force field is parametrized against fluid-phase equilibria data with a strong emphasis on transferability. The term *transferable* implies that the same force field parameters are used to describe a given interaction site in different molecules (e.g., identical parameters should be used for the methyl group in *n*-pentane, 1-pentene, and 1-pentanol) and that the force field is applicable to predict different properties (e.g., thermodynamic, structural, or transport) across a wide range of state points (e.g., pressure, temperature, or composition).

Four major versions of the force fields exist for (mostly) organic molecules. They differ in sophistication: TraPPE-CG (coarse grain), TraPPE-UA (united-atom), TraPPE-EH (explicit-hydrogen), and TraPPE-pol (polarizable). Further, TraPPE-SM (small molecule) and TraPPE-zeo (zeolites) covers CO_2, N_2, O_2, NH_3, zeolites, etc. As of 2016, parts of the TraPPE force field are implemented in several software packages including To-whee, Materials Design, Culgi, and Scinomics.

Functional Form

The basic functional form of the TraPPE force field is (for the united-atom version):

$$U(r^N) = \sum_{j=1}^{N-1} \sum_{i=j+1}^{N} \{4\varepsilon_{ij}[\left(\frac{\sigma_{ij}}{r_{ij}}\right)^{12} - \left(\frac{\sigma_{ij}}{r_{ij}}\right)^{6}] + \frac{q_i q_j}{4\pi\varepsilon_0 r_{ij}}\} + \sum_{angles} \frac{k_a(\theta - \theta_0)^2}{2} + U_{torsion}$$

Some considerations regarding the model:

- In the united-atom model, a CH_x group is treated as one interaction site or *pseudo atom* located on the carbon center.

- TraPPE typically uses fixed bond lengths and thus does not include a bond stretching term in the potential. However, the molecule is still semi-flexible due to the bending and torsional degrees of freedom.

- The double summation over site indices i and j represents nonbonded interactions between two pseudo atoms of different molecules or of the same molecule but separated by (usually) at least four bonds.

- Lennard-Jones potential (first term of summation) is used to describe repulsion and dispersion. σ_{ij} is related to the equilibrium distance, $R_{0,ij}$, by: $\sigma_{ij} = R_{0,ij} / 2^{1/6}$ and ε_{ij} is the well depth. For unlike Lennard-Jones interactions, standard Lorentz–Berthelot combining rules are used.

- Coulomb or electric potential (second term of summation) is used to describe first-order electrostatic interactions.

- The parameters for the Lennard-Jones and Coulomb potentials reflect effective values that account in a mean-field manner for higher-order and many-body dispersion and induction effects. In general, the parameters used in the TraPPE force field are fit to the vapor liquid coexistence curves of a few selected target compounds, but are found to reproduce transport properties also.

Parameter Set

The parameters for the TraPPE force field can be obtained from the TraPPE website.

References

- Hill, Terrell L. (1946). "On Steric Effects". The Journal of Chemical Physics. 14 (7): 465–465. ISSN 0021-9606. doi:10.1063/1.1724172

- Merck molecular force field. I. Basis, form, scope, parameterization, and performance of MMFF94, Thomas A. Halgren, J. Comp. Chem.; 1996; 490-519, doi:10.1002/(SICI)1096-987X(199604)17:5/6<490::AID-JCC1>3.0.CO;2-P

- Kearsley, Simon (June 1999). "MMFF94 Validation Suite". CCL.net. Computational Chemistry List, Ltd. Retrieved 17 September 2016

- Marrink, Siewert J.; de Vries, Alex H.; Mark, Alan E. (1 January 2004). "Coarse Grained Model for Semiquantitative Lipid Simulations". The Journal of Physical Chemistry B. 108 (2): 750–760. doi:10.1021/jp036508g

- Karplus M (2006). "Spinach on the ceiling: a theoretical chemist's return to biology". Annu Rev Biophys Biomol Struct. 35 (1): 1–47. PMID 16689626. doi:10.1146/annurev.biophys.33.110502.133350

- "TraPPE: Transferable Potentials for Phase Equilibria". The Siepmann Group. University of Minnesota. Retrieved February 4, 2016

- Deetz, J. D.; Faller, R. (2014). "Parallel Optimization of a Reactive Force Field for Polycondensation of Alkoxysilanes". The Journal of Physical Chemistry B. 118 (37): 10966. doi:10.1021/jp504138r

- Brooks BR, Bruccoleri RE, Olafson BD, States DJ, Swaminathan S, Karplus M (1983). "CHARMM: A program for macromolecular energy, minimization, and dynamics calculations". J Comp Chem. 4 (2): 187–217. doi:10.1002/jcc.540040211

Softwares Used in Computational Chemistry

Some of the software used in computational chemistry include the Advanced Simulation Library (ASL), Crystal, Gaussian, etc. Gaussian can perform functions pertaining to AMBER, Universal Force Field (UFF), Hartree-Fock method among others. The topics discussed in the chapter are of great importance to broaden the existing knowledge on computational chemistry.

TeraChem

TeraChem is the first computational chemistry software program written completely from scratch to benefit from the new streaming processors such as graphics processing units (GPUs). The computational algorithms have been completely redesigned to exploit massive parallelism of CUDA-enabled Nvidia GPUs. The original development started at the University of Illinois at Urbana-Champaign. Due to the great potential of the developed technology, this GPU-accelerated software was subsequently commercialized. Now it is distributed by PetaChem, LLC, located in the Silicon Valley. The software package is under active development and new features are released often.

Core Features

Very fast *ab initio* molecular dynamics and density functional theory (DFT) methods for *nanoscale* biomolecular systems with hundreds of atoms are arguably the most attractive features of TeraChem. Its affinity to extreme performance is also exemplified in the TeraChem motto *"Chemistry at the Speed of Graphics"*. All the methods used are based on Gaussian orbitals, a choice made to improve performance on the limited computing capacities of modern computer hardware.

Press Coverage

- Chemical and Engineering News (C&EN) magazine of the American Chemical Society first mentioned the development of TeraChem in one of its Fall 2008 issues. Then, GPU-accelerated computing was at the level of a very extravagant science.

- Recently, C&EN magazine has a feature article covering molecular modeling on GPU and TeraChem.

- According to the recent post at the Nvidia blog, TeraChem has been tested to deliver 8-50 times better performance than General Atomic and Molecular Structure System (GAMESS). In that benchmark, TeraChem was executed on a *desktop* machine with four (4) Tesla GPUs and GAMESS was running on a cluster of 256 quad core CPUs.

- TeraChem is available for free via GPU Test Drive.

Media

The software is featured in a series of clips on its own YouTube channel under "GPU-Chem" user.

- TeraChem v1.5 release link

- New kinds of science enabled: ab initio dynamics of proton transfer link

- Discovery mode: reactions in nanocavities link

- TeraChem performance on 4 GPUs: video

Major Release History

2016

- TeraChem version 1.9

  Support for Maxwell cards (e.g., GTX980, TitanX)

  Effective core potentials (and gradients)

  Time-dependent density functional theory

  Continuum solvation models (COSMO)

2012

- TeraChem version 1.5

  Full support of polarization functions: energy, gradients, ab initio dynamics and range-corrected DFT functionals (CAMB3LYP, wPBE, wB97x)

2011

- TeraChem version 1.5a (pre-release)

  Alpha version with the full support of d-functions: energy, gradients, ab initio dynamics

- TeraChem version 1.43b-1.45b

  Beta version with polarization functions for energy calculation (HF/DFT levels) as well as other improvements.

- TeraChem version 1.42

  This version was first deployed at National Center for Supercomputing Applications' (NCSA) Lincoln supercomputer for National Science Foundation (NSF) TeraGrid users as announced in NCSA press release.

2010

- TeraChem version 1.0

- TeraChem version 1.0b

  The very first initial beta release was reportedly downloaded more than 4,000 times.

Advanced Simulation Library

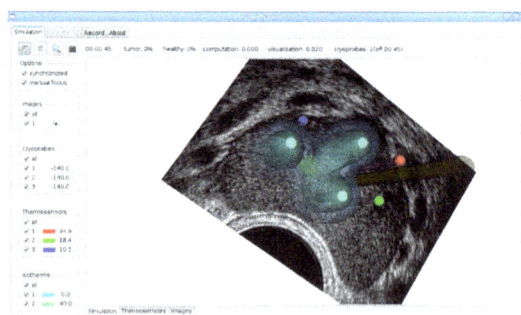

Computer-assisted cryosurgery

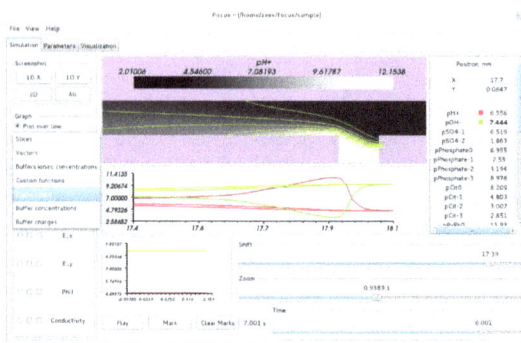

Simulation of a microfluidic device for separating mixtures of proteins
Coating procedure employing physical vapor deposition (PVD) method.

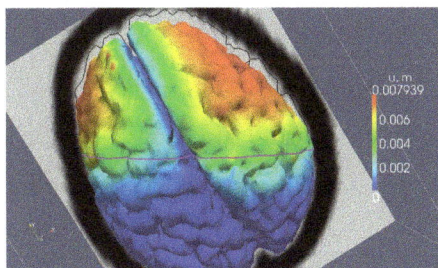

Image-guided neurosurgery, brain deformation simulation

Aerodynamics of a locomotive in a tunnel

Advanced Simulation Library (ASL) is free and open-source hardware-accelerated multiphysics simulation platform. It enables users to write customized numerical solvers in C++ and deploy them on a variety of massively parallel architectures, ranging from inexpensive FPGAs, DSPs and GPUs up to heterogeneous clusters and supercomputers. Its internal computational engine is written in OpenCL and utilizes matrix-free solution techniques. ASL implements variety of modern numerical methods, i.a. level-set method, lattice Boltzmann, immersed Boundary. Mesh-free, immersed boundary approach allows to move from CAD directly to simulation, reducing pre-processing efforts and amount of potential errors. ASL can be used to model various coupled physical and chemical phenomena, especially in the field of computational fluid dynamics. It is distributed under the free GNU Affero General Public License with an optional commercial license (which is based on the permissive MIT License).

History

Advanced Simulation Library is being developed by Avtech Scientific, an Israeli company. Its source code was released to the community on 14 May 2015, whose members packaged it for scientific sections of all major Linux distributions shortly thereafter. Subsequently, Khronos Group acknowledged the significance of ASL and listed it on its website among OpenCL-based resources.

Application Areas

- Computational fluid dynamics

- Computer-assisted surgery

- Virtual sensing

- Industrial process data validation and reconciliation

- Multidisciplinary design optimization

- Design space exploration

- Computer-aided engineering

- Crystallography

- Microfluidics

Advantages and Disadvantages

Advantages

- C++ API (no OpenCL knowledge required)

- Mesh-free, immersed boundary approach allows to move from CAD directly to computations reducing pre-processing effort

- Dynamic compilation enables an additional layer of optimization at run-time (i.e. for a specific parameters set the application was provided with)

- Automatic hardware acceleration and parallelization of applications

- Deployment of same program on a variety of parallel architectures - GPU, APU, FPGA, DSP, multicore CPUs

- Ability to deal with complex boundaries

- Ability to incorporate microscopic interactions

- Availability of the source code

Disadvantages

- Absence of detailed documentation (besides the Developer Guide generated from the source code comments)

- Not all OpenCL drivers are mature enough for the library

Features

ASL provides a range of features to solve number of problems - from complex fluid flows involving chemical reactions, turbulence and heat transfer, to solid mechanics and elasticity.

- Interfacing: VTK/ParaView, MATLAB (export).

  o import file formats: .stl .vtp .vtk .vti .mnc .dcm

  o export file formats: .vti .mat

- Geometry:

  o flexible and complex geometry using simple rectangular grid

  o mesh-free, immersed boundary approach

  o generation and manipulation of geometric primitives

- Implemented phenomena:

  o Transport processes

    ▪ multicomponent transport processes

    ▪ compressible and incompressible fluid flow

  o Chemical reactions

    ▪ electrode reactions

  o Elasticity

    ▪ homogeneous isotropic elasticity

    ▪ homogeneous isotropic poroelasticity

  o Interface tracking

    ▪ evolution of an interface

    ▪ evolution of an interface with crystallographic kinetics

Uses

- ACTIVE - Active Constraints Technologies for Ill-defined or Volatile Environments (European FP7 Project).

Crystal (Software)

CRYSTAL is a quantum chemistry ab initio program, designed primarily for calculations on crystals (3 dimensions), slabs (2 dimensions) and polymers (1 dimension) using translational symmetry, but it can also be used for single molecules. It is written by V.R. Saunders, R. Dovesi, C. Roetti, R. Orlando, C.M. Zicovich-Wilson, N.M. Harrison, K. Doll, B. Civalleri, I.J. Bush, Ph. D'Arco, and M. Llunell from Theoretical Chemistry

Group at the University of Torino and the Computational Materials Science Group at the Daresbury Laboratory near Warrington in Cheshire, England. The current version is CRYSTAL14, released in June 2014. Earlier versions were CRYSTAL88, CRYSTAL92, CRYSTAL95, CRYSTAL98, CRYSTAL03, CRYSTAL06, and CRYSTAL09.

Program Structure

The program is built of two modules: *crystal* and *properties*. The *crystal* program is dedicated to perform the SCF calculations, the geometry optimizations, and the frequency calculations for the structures given in input. At the end of the SCF process, the program crystal writes information on the crystalline system and its wave function as unformatted sequential data in Fortran unit 9, and as formatted data in Fortran unit 98. One-electron properties and wave function analysis can be computed from the SCF wave function by running the program *properties*.

The main advantage of the crystal code is due to the deep and optimized exploitation of symmetry, at all levels of calculation (SCF as well gradients and vibrational frequencies calculations). This allows significant reduction of the computational cost for periodic calculations. Note that while the symmetry generally reduces to identity in large molecules, large crystalline system usually show many symmetry operators.

Gaussian (Software)

Gaussian is a general purpose computational chemistry software package initially released in 1970 by John Pople and his research group at Carnegie Mellon University as Gaussian 70. It has been continuously updated since then. The name originates from Pople's use of Gaussian orbitals to speed up molecular electronic structure calculations as opposed to using Slater-type orbitals, a choice made to improve performance on the limited computing capacities of then-current computer hardware for Hartree–Fock calculations. The current version of the program is Gaussian 16. Originally available through the Quantum Chemistry Program Exchange, it was later licensed out of Carnegie Mellon University, and since 1987 has been developed and licensed by Gaussian, Inc.

Gaussian quickly became a popular and widely used electronic structure program. Prof. Pople and his students and post-docs were among those who pushed the development of the package, including cutting-edge research in quantum chemistry and other fields.

Standard Abilities

According to the most recent Gaussian manual, the package can do:

- Molecular mechanics

- o AMBER

- o Universal force field (UFF)

- o DREIDING force field

- Semi-empirical quantum chemistry method calculations

  - o Austin Model 1 (AM1), PM3, CNDO, INDO, MINDO/3, MNDO

- Self-consistent field (SCF methods)

  - o Hartree–Fock method: restricted, unrestricted, and restricted open-shell

- Møller–Plesset perturbation theory (MP2, MP3, MP4, MP5).

- Built-in density functional theory (DFT) methods

  - o B3LYP and other hybrid functionals

  - o Exchange functionals: PBE, MPW, PW91, Slater, X-alpha, Gill96, TPSS.

  - o Correlation functionals: PBE, TPSS, VWN, PW91, LYP, PL, P86, B95

- ONIOM (QM/MM method) up to three layers

- Complete active space (CAS) and multi-configurational self-consistent field calculations

- Coupled cluster calculations

- Quadratic configuration interaction (QCI) methods

- Quantum chemistry composite methods – CBS-QB3, CBS-4, CBS-Q, CBS-Q/APNO, G1, G2, G3, W1 high-accuracy methods.

Official Release History

Gaussian 70, Gaussian 76, Gaussian 80, Gaussian 82, Gaussian 86, Gaussian 88, Gaussian 90, Gaussian 92, Gaussian 92/DFT, Gaussian 94, Gaussian 98, Gaussian 03, Gaussian 09 and Gaussian 16.

Other programs named 'Gaussian XX' were placed among the holdings of the Quantum Chemistry Program Exchange. These were unofficial, unverified ports of the program to other computer platforms.

License Controversy

In the past, Gaussian, Inc. has attracted controversy for its licensing terms that stipulate that researchers who develop competing software packages are not permitted to use the software. Some scientists consider these terms overly restrictive. The anonymous group *bannedbygaussian.org* has published a list of scientists whom it claims are not permitted to use GAUSSIAN software. These assertions were repeated by Jim Giles in 2004 in *Nature*. The controversy was also noted in 1999 by *Chemical and Engineering News* (repeated without additional content in 2004), and in 2000, the World Association of Theoretically Oriented Chemists Scientific Board held a referendum of its executive board members on this issue with a majority (23 of 28) approving the resolution opposing the restrictive licenses.

Gaussian, Inc. disputes the accuracy of these descriptions of its policy and actions, noting that all of the listed institutions do in fact have licenses for everyone but directly competing researchers. They also claim that not licensing competitors is standard practice in the software industry and members of the Gaussian collaboration community have been refused licenses from competing institutions.

Columbus

Columbus is a computational chemistry software suite for calculating ab initio molecular electronic structures, designed as a collection of individual programs communicating through files. The programs focus on extended multi-reference calculations of atomic and molecular ground and excited states. Besides standard classes of reference wave functions such as CAS and RAS, calculations can be performed with selected configurations. It makes use of the atomic orbital integrals and gradient routines from the DALTON program. The program is available free of charge under license (including the DALTON license).

COLUMBUS is frequently used for nonadiabatic problems because of its ability to calculate MRCI nonadiabatic coupling vector analytically.

Brief History

COLUMBUS was started in 1980 in the Department of Chemistry of the Ohio State University by Isaiah Shavitt, Hans Lischka and Ron Shepard. The program pioneered the Graphical Unitary Group Approach (GUGA) for configuration interaction calculations, which is now available in many other program suites. The program is named after Columbus, OH.

Style

COLUMBUS maintains a program unique style that distinguish itself from most other quantum chemistry programs.

The program suite is a collection of a number of programs coded in Fortran, each can be executed independently. These programs communicate through files. Perl scripts are provided to prepare input files and to link these programs together to perform common tasks such as single point energy calculation, geometry optimization, normal mode analysis, etc. This style provides very high degree of flexibility which is embraced by advanced users. The open style allows new components to be added to the program suite with ease. However, such flexibility also increased the complexity of input file preparation and execution, making it very difficult for new users.

Major Features

- Hartree–Fock method (closed-shell and restricted open-shell).

- Multi-configurational self-consistent field (MCSCF) (quadratic convergence and state averaging).

- Multi-reference CISD for an arbitrary set of reference configurations (including a massively parallel version).

- Configuration interaction calculations are based on Graphical Unitary Group Approach (GUGA).

- Analytic gradients for MCSCF, MR-CISD, MR-ACPF and MR-AQCC.

- Analytic MCSCF and MR-CISD nonadiabatic coupling vectors.

- Support for electrostatic embedding QM/MM calculations.

- Automatic geometry optimization, saddle-point searches.

- Automatic searches for the minima on conical intersection seams.

- Spin/orbit configuration interaction.

Octopus (Software)

Octopus is a software package for performing Kohn–Sham density functional theory (DFT) and time-dependent density functional theory (TDDFT) calculations.

octopus employs pseudopotentials and real-space numerical grids to propagate the Kohn–Sham orbitals in real time under the influence of time-varying electromagnetic fields. Specific functionality is provided for simulating one-, two-, and three-dimensional systems. octopus can calculate static and dynamic polarizabilities and first hyperpolarizabilities, static magnetic susceptibilities, absorption spectra, and perform molecular dynamics simulations with Ehrenfest and Car–Parrinello methods.

The code is written predominantly in Fortran, with some C and Perl. It is released under the GPL.

Target Problems

- Linear optical (i.e. electronic) response of molecules or clusters, also second-order nonlinear response.

- Non-linear response to classical high-intensity electromagnetic fields, taking into account both the ionic and electronic degrees of freedom.

- Ground-state and excited state electronic properties of systems with lower dimensionality, such as quantum dots.

- Photo-induced reactions of molecules (e.g., photo-dissociation, photo-isomerization, etc.).

- In the immediate future, extension of these procedures to systems that are infinite and periodic in one or more dimensions (polymers, slabs, nanotubes, solids), and to electronic transport.

Theoretical Basis

- The underlying theories are DFT and TDDFT. Also, the code may perform dynamics by considering the classical (i.e. point-particle) approximation for the nuclei. These dynamics may be non-adiabatic, since the system evolves following the Ehrenfest path. It is, however, a mean-field approach.

- Regarding TDDFT, one can use three different approaches:

  o the standard TDDFT-based linear-response theory of Casida, which provides the excitation energies and oscillator strengths for ground-state to excited-state transitions.

  o the explicit time-propagation of the TDDFT equations, which allows for the use of large external potentials, well beyond the range of validity of perturbation theory.

  o the Sternheimer equation (density-functional perturbation theory) in the frequency domain, using only occupied states.

Methodology

- As numerical representation, the code works without a basis set, relying on numerical meshes. Nevertheless, auxiliary basis sets (plane waves, atomic orbitals) are used when necessary. Recently, the code offers the possibility of working with non-uniform grids, which adapt to the inhomogeneity of the problem, and of making use of multigrid techniques to accelerate the calculations.

- For most calculations, the code relies on the use of pseudopotentials of two types: Troullier-Martins, and Hartwigsen-Goedecker-Hutter.

- In addition to being able to treat systems in the standard 3 dimensions, 2D and 1D modes are also available. These are useful for studying, e.g., the two-dimensional electron gas that characterizes a wide class of quantum dots.

Technical Aspects

- The code has been designed with emphasis on parallel scalability. In consequence, it allows for multiple task divisions, this utilises mesh division software, MPI and OpenMP.

- The language of most of the code is Fortran 90 (almost 50.000 lines at present). Other languages, such as C or Perl, are also used.

- The package is licensed under the GNU General Public License (GPL). In consequence, it is available for use, inspection, and modification for anyone, at the octopus web page.

Spartan (Software)

Spartan is a molecular modelling and computational chemistry application from Wavefunction. It contains code for molecular mechanics, semi-empirical methods, *ab initio* models, density functional models, post-Hartree–Fock models, and thermochemical recipes including G3(MP2) and T1.

Primary functions are to supply information about structures, relative stabilities and other properties of isolated molecules. Molecular mechanics calculations on complex molecules are common in the chemical community. Quantum chemical calculations, including Hartree–Fock method molecular orbital calculations, but especially calculations that include electronic correlation, are more time consuming in comparison.

Quantum chemical calculations are also called upon to furnish information about mechanisms and product distributions of chemical reactions, either directly by calculations on transition states, or based on Hammond's postulate, by modeling the steric and electronic demands of the reactants. Quantitative calculations, leading directly to information about the geometries of transition states, and about reaction mechanisms in general, are increasingly common, while qualitative models are still needed for systems that are too large to be subjected to more rigorous treatments. Quantum chemical calculations can supply information to complement existing experimental data or replace it altogether, for example, atomic charges for quantitative structure-activity relationship (QSAR) analyses, and intermolecular potentials for molecular mechanics and molecular dynamics calculations.

Spartan applies computational chemistry methods (theoretical models) to many standard tasks that provide calculated data applicable to the determination of molecular shape conformation, structure (equilibrium and transition state geometry), NMR, IR, Raman, and UV-visible spectra, molecular (and atomic) properties, reactivity, and selectivity.

Computational Abilities

This software provides the molecular mechanics, Merck Molecular Force Field (MMFF), (for validation test suite), MMFF with extensions, and SYBYL, force fields calculation, Semi-empirical calculations, MNDO/MNDO(D), Austin Model 1 (AM1), PM3, Recife Model 1 (RM1) PM6.

- *Hartree–Fock, self-consistent field (SCF) methods*, available with implicit solvent (SM8).

    o   Restricted, unrestricted, and restricted open-shell Hartree–Fock

- *Density functional theory (DFT) methods*, available with implicit solvent (SM8).

    o   *Standard functionals*: BP, BLYP, B3LYP, EDF1, EDF2, M06, ωB97X-D

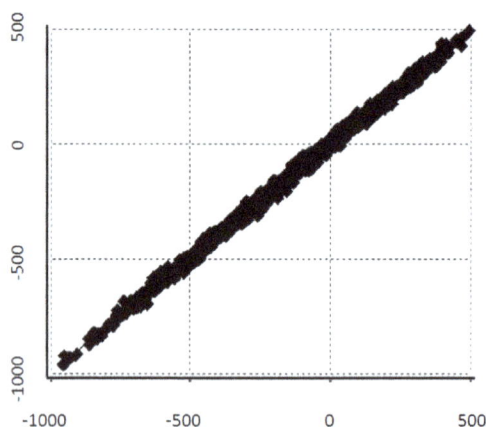

The calculated T1 heat of formation (y axis) relative to the experimental heat of formation (x axis) for a set of >1800 diverse organic molecules from the NIST thermochemical database with mean absolute and RMS errors of 8.5 and 11.5 kJ/mol, respectively.

    o   *Exchange functionals*: HF, Slater-Dirac, Becke88, Gill96, GG99, B(EDF1), PW91

    o   *Correlation functionals*: VWN, LYP, PW91, P86, PZ81, PBE.

    o   *Combination or hybrid functionals*: B3PW91, B3LYP, B3LYP5, EDF1, EDF2, BMK

- Truhlar group functionals: M05, M05-2X, M06, M06-L M06-2X, M06-HF

- Head-Gordon group functionals: ωB97, ωB97X, ωB97X-D

- *Coupled cluster methods.*

  - CCSD, CCSD(T), CCSD(2), OD, OD(T), OD(2), QCCD, VOD, VOD(2), VQC-CD

- *Møller–Plesset methods.*

  - MP2, MP3, MP4, RI-MP2

- *Excited state methods.*

  - *Time-dependent density functional theory (TDDFT)*

  - *Configuration interaction*: CIS, CIS(D), QCIS(D), quadratic configuration interaction (QCISD(T)), RI-CIS(D)

- *Quantum chemistry composite methods, thermochemical recipes.*

  - T1, G2, G3, G3(MP2)

Tasks Performed

Available computational models provide molecular, thermodynamic, QSAR, atomic, graphical, and spectral properties. A calculation dialogue provides access to the following computational tasks:

- Energy - For a given geometry, provides energy and associated properties of a molecule or system. If quantum chemical models are employed, the wave function is calculated.

- Equilibrium molecular geometry - Locates the nearest local minimum and provides energy and associated properties.

- Transition state geometry - Locates the nearest first-order saddle point (a maximum in a single dimension and minima in all others) and provides energy and associated properties.

- Equilibrium conformer - Locates lowest-energy conformation. Often performed before calculating structure using a quantum chemical model.

- Conformer distribution - Obtains a selection of low-energy conformers. Commonly used to identify the shapes a specific molecule is likely to adopt and to determine a Boltzmann distribution for calculating average molecular properties.

- Conformer library - Locates lowest-energy conformer and associates this with a set of conformers spanning all shapes accessible to the molecule without regard to energy. Used to build libraries for similarity analysis.

- Energy profile - Steps a molecule or system through a user defined coordinate set, providing equilibrium geometries for each step (subject to user-specified constraints).

- Similarity analysis - quantifies the likeness of molecules (and optionally their conformers) based on either structure or chemical function (Hydrogen bond acceptors–donors, positive–negative ionizables, hydrophobes, aromatics). Quantifies likeness of a molecule (and optionally its conformers) to a pharmacophore.

Graphical User Interface

The software contains an integrated graphical user interface. Touch screen operations are supported for Windows 7 and 8 devices. Construction of molecules in 3D is facilitated with molecule builders (included are organic, inorganic, peptide, nucleotide, and substituent builders). 2D construction is supported for organic molecules with a 2D sketch palette. The Windows version interface can access ChemDraw; which versions 9.0 or later may also be used for molecule building in 2D. A calculations dialogue is used for specification of task and computational method. Data from calculations are displayed in dialogues, or as text output. Additional data analysis, including linear regression, is possible from an internal spreadsheet.

Graphical Models

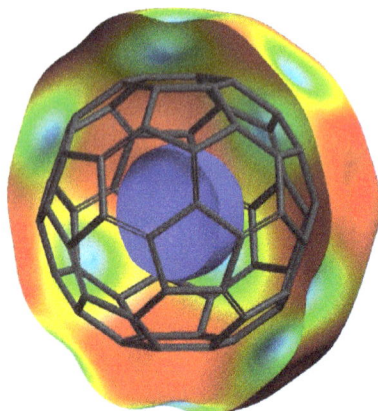

A cut-away view of the *electrostatic potential map* of fullerene (C_{60}), the blue area inside the molecule is an area of positive charge (relative to the superstructure, providing a pictorial explanation for fullerene's ability to encapsulate negatively charged species).

Graphical models, especially molecular orbitals, electron density, and electrostatic potential maps, are a routine means of molecular visualization in chemistry education.

- *Surfaces*:

    o   Molecular orbitals (highest occupied, lowest unoccupied, and others)

    o   Electron density - The density, $\rho(r)$, is a function of the coordinates r, defined such that $\rho(r)dr$ is the number of electrons inside a small volume dr. This is what is measured in an X-ray diffraction experiment. The density may be portrayed in terms of an isosurface (isodensity surface) with the size and shape of the surface being given by the value (or percentage of enclosure) of the electron density.

    o   Spin density - The density, $\rho^{spin}(r)$, is defined as the difference in electron density formed by electrons of α spin, $\rho\alpha(r)$, and the electron density formed by electrons of β spin, $\rho\beta(r)$. For closed-shell molecules (in which all electrons are paired), the spin density is zero everywhere. For open-shell molecules (in which one or more electrons are unpaired), the spin density indicates the distribution of unpaired electrons. Spin density is an indicator of reactivity of radicals.

    o   Van der Waals radius (surface)

    o   Solvent accessible surface area

    o   Electrostatic potential - The potential, ε_p, is defined as the energy of interaction of a positive point charge located at p with the nuclei and electrons of a molecule. A surface for which the electrostatic potential is negative (a negative potential surface) delineates regions in a molecule which are subject to electrophilic attack.

- *Composite surfaces (maps)*:

    o   Electrostatic potential map (electrophilic indicator) - The most commonly employed property map is the electrostatic potential map. This gives the potential at locations on a particular surface, most commonly a surface of electron density corresponding to overall molecular size.

    o   Local ionization potential map - Is defined as the sum over orbital electron densities, $\rho i(r)$ times absolute orbital energies, εi, and divided by the total electron density, $\rho(r)$. The local ionization potential reflects the relative ease of electron removal ("ionization") at any location around a molecule. For example, a surface of "low" local ionization potential for sulfur tetrafluoride demarks the areas which are most easily ionized.

    o   LUMO map (nucleophilic indicator) - Maps of molecular orbitals may also lead to graphical indicators. For example, the *LUMO map*, wherein the (absolute value) of the lowest-unoccupied molecular orbital (the LUMO) is mapped onto a size surface (again, most commonly the electron density), providing an indication of nucleophilic reactivity.

Spectral Calculations

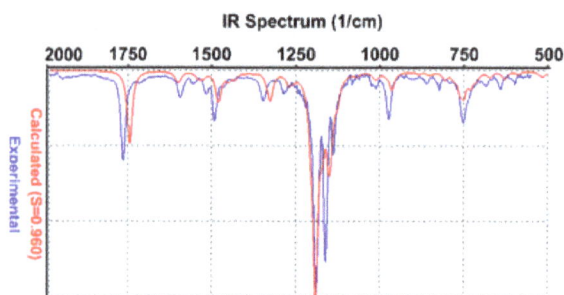

The calculated (DFT/EDF2/6-31G*) IR spectra (red), scaled and optimized to the experimental FT-IR spectra (blue) of phenyl 9-acridinecarboxylate (below).

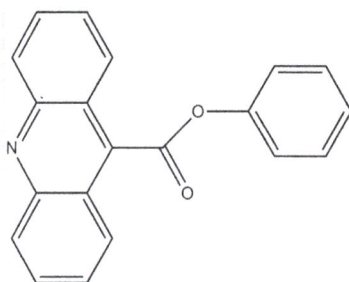

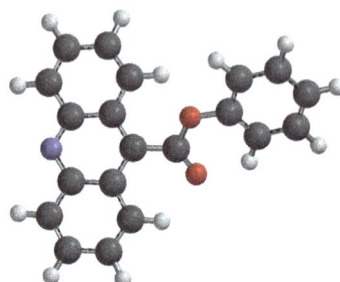

2D rendering   3D rendering

The molecule phenyl 9-acridinecarboxylate.

Available spectra data and plots for:

- *Infrared spectroscopy (IR) spectra*

  o Fourier transform spectroscopy (FT-IR)

  o Raman spectroscopy (IR)

- *Nuclear magnetic resonance (NMR) spectra*

  o ^{1}H chemical shifts and coupling constants (empirical)

  o ^{13}C chemical shifts, Boltzmann averaged shifts, and ^{13}C DEPT spectra

  o 2D H vs H Spectra

    ▪ COSY plots

  o 2D C vs H Spectra

    ▪ Heteronuclear single-quantum correlation spectroscopy (HSQC) spectra

    ▪ HMBC spectra

- *UV/vis Spectra*

Experimental spectra may be imported for comparison with calculated spectra: IR and UV/vis spectra in Joint Committee on Atomic and Molecular Physical Data (JCAMP) (.dx) format and NMR spectra in Chemical Markup Language (.cml) format. Access to public domain spectral databases is available for IR, NMR, and UV/vis spectra.

Databases

Spartan accesses several external databases.

- *Quantum chemical calculations databases:*

  o Spartan Spectra & Properties Database (SSPD) - a set of about 252,000 molecules, with structures, energies, NMR and IR spectra, and wave functions calculated using the EDF2 density functional theory with the 6-31G* basis set.

  o Spartan Molecular Database (SMD) - a set of about 100,000 molecules calculated from following models:

    - Hartree–Fock with 3-21G, 6-31G*, and 6-311+G** basis sets

    - B3LYP density functional with 6-31G* and 6-311+G** basis sets

    - EDF1 density functional with 6-31G* basis set

    - MP2 with 6-31G* and 6-311+G** basis sets

    - G3(MP2)

    - T1

- *Experimental databases:*

  o NMRShiftDB - an open-source database of experimental ^{1}H and ^{13}C chemical shifts.

  o Cambridge Structural Database (CSD) - a large repository of small molecule organic and inorganic experimental crystal structures of about 600,000 entries.

  o NIST database of experimental IR and UV/vis spectra.

Major Release History

- 1991 Spartan version 1 Unix

- 1993 Spartan version 2 Unix

- 1994 Mac Spartan Macintosh

- 1995 Spartan version 3 Unix

- 1995 PC Spartan Windows

- 1996 Mac Spartan Plus Macintosh

- 1997 Spartan version 4 Unix

- 1997 PC Spartan Plus Windows

- 1999 Spartan version 5 Unix

- 1999 PC Spartan Pro Windows

- 2000 Mac Spartan Pro Macintosh

- 2002 Spartan'02 Unix, Linux, Windows, Mac

Windows, Macintosh, Linux Versions

- 2004 Spartan'04

- 2006 Spartan'06

- 2008 Spartan'08

- 2010 Spartan'10

- 2013 Spartan'14

- 2016 Spartan'16

ACES (Computational Chemistry)

Aces II (Advanced Concepts in Electronic Structure Theory) is an ab initio computational chemistry package for performing high-level quantum chemical ab initio calculations. Its major strength is the accurate calculation of atomic and molecular energies as well as properties using many-body techniques such as many-body perturbation theory (MBPT) and, in particular coupled cluster techniques to treat electron correlation. The development of ACES II began in early 1990 in the group of Professor Rodney J. Bartlett at the Quantum Theory Project (QTP) of the University of Florida in Gainesville. There, the need for more efficient codes had been realized and the idea of writing an entirely new program package emerged. During 1990 and 1991 John F. Stanton, Jürgen Gauß, and John D. Watts, all of them at that time postdoctoral researchers in the Bartlett group, supported by a few students, wrote the backbone of what is now known as the ACES II program package. The only parts which were not new coding efforts were the integral packages (the MOLECULE package of J. Almlöf, the VPROP package of P.R. Taylor, and the integral derivative package ABACUS of T. Helgaker, P. Jorgensen J. Olsen, and H.J. Aa. Jensen). The latter was modified extensively for adaptation with Aces II, while the others remained very much in their original forms.

Ultimately, two different versions of the program evolved. The first was maintained by the Bartlett group at the University of Florida, and the other (known as ACE-SII-MAB) was maintained by groups at the University of Texas, Universitaet Mainz in Germany, and ELTE in Budapest, Hungary. The latter has recently been renamed as CFOUR.

Aces III is a parallel implementation that was released in the fall of 2008. The effort led to definition of a new architecture for scalable parallel software called the super instruction architecture. The design and creation of software is divided into two parts:

1. The algorithms are coded in a domain specific language called super instruction assembly language or SIAL, pronounced "sail" for easy communication.

2. The SIAL programs are executed by a MPMD parallel virtual machine called the super instruction processor or SIP.

The ACES III program consists of 580,000 lines of SIAL code of which 200,000 lines are comments, and 230,000 lines of C/C++ and Fortran of which 62,000 lines are comments.

Jaguar (Software)

Jaguar is an ab *initio* quantum chemistry package for both gas and solution phase calculations, with strength in treating metal-containing systems. It is commercial software marketed by the company Schrödinger. The program was originated in research groups of Richard Friesner and William Goddard and was initially called PS-GVB (referring to the so-called pseudospectral generalized valence bond method that the program featured).

Jaguar is an essential component of two other Schrödinger products. The program Maestro provides the graphical user interface to Jaguar, and a QM/MM program QSite uses Jaguar as its quantum-chemical engine.

Features

A distinctive feature of Jaguar is its use of the pseudospectral approximation. This approximation can be applied to computationally expensive integral operations present in most quantum chemical calculations. As a result, calculations complete much faster at a negligible loss of accuracy.

The current version Jaguar 8.0 includes the following capabilities:

- Hartree–Fock (RHF, UHF, ROHF) and density functional theory (LDA, gradient-corrected, dispersion-corrected, and hybrid functionals)

- Local second-order Møller–Plesset perturbation theory (LMP2)

- Generalized valence bond perfect-pairing (GVB-PP) and GVB-LMP2 calculations

- Prediction of excited states using configuration interaction (CIS) and time-dependent density functional theory (TDDFT)

- Geometry optimization and transition state search

- Solvation calculations based on the Poisson–Boltzmann equation

- Prediction of infrared (IR), nuclear magnetic resonance (NMR), ultraviolet (UV), and vibrational circular dichroism (VCD) spectra

- PKa prediction

- Generation of various molecular surfaces (electrostatic potential, electron density, molecular orbitals etc.)

- Prediction of various molecular properties (multipole moments, polarizabilities, vibrational frequencies etc.)

Known Version History

• Jaguar 8.0 (2013)	• Jaguar 6.0 (2005)
• Jaguar 7.9 (2012)	• Jaguar 5.5 (2004)
• Jaguar 7.8 (2011)	• Jaguar 5.0 (2003)
• Jaguar 7.7 (2010)	• Jaguar 4.2 (2002)
• Jaguar 7.6 (2009)	• Jaguar 4.1 (2001)
• Jaguar 7.5 (2008)	• Jaguar 4.0 (2000)
• Jaguar 7.0 (2007)	• Jaguar 3.5
• Jaguar 6.5 (2006)	• Jaguar 3.0

Desmond (Software)

Desmond is a software package developed at D. E. Shaw Research to perform high-speed molecular dynamics simulations of biological systems on conventional computer clusters. The code uses novel parallel algorithms and numerical methods to achieve high performance on platforms containing a large number of processors, but may also be executed on a single computer.

The core and source code are available at no cost for non-commercial use by universities and other not-for-profit research institutions, and have been used in the Folding@home distributed computing project. Desmond is available as commercial software through Schrödinger, Inc.

Molecular Dynamics Program

Desmond supports algorithms typically used to perform fast and accurate molecular dynamics. Long-range electrostatic energy and forces can be calculated using particle mesh Ewald-based methods. Constraints can be enforced using the M-SHAKE algorithm. These methods can be used together with time-scale splitting (RESPA-based) integration schemes.

Desmond can compute energies and forces for many standard fixed-charged force fields used in biomolecular simulations, and is also compatible with polarizable force fields based on the Drude formalism. A variety of integrators and support for various ensembles have been implemented in the code, including methods for temperature control (Andersen, Nosé-Hoover, and Langevin) and pressure control (Berendsen, Martyna-Tobias-Klein, and Langevin). The code also supports methods for restraining atomic positions and molecular configurations; allows simulations to be carried out using a variety of periodic cell configurations; and has facilities for accurate check-pointing and restart.

Desmond can also be used to perform absolute and relative free energy calculations (e.g., free energy perturbation). Other simulation methods (such as replica exchange) are supported through a plug-in-based infrastructure, which also allows users to develop their own simulation algorithms and models.

Desmond is also available in a graphics processing unit (GPU) accelerated version that is about 60-80 times faster than the central processing unit (CPU) version.

Related Software Tools

Along with the molecular dynamics program, the Desmond software also includes tools for minimizing and energy analysis, both of which can be run efficiently in a parallel environment.

Force fields parameters can be assigned using a template-based parameter assignment tool called Viparr. It currently supports several versions of the CHARMM, Amber and OPLS force fields, and a range of different water models.

Desmond is integrated with a molecular modeling environment (Maestro, developed by Schrödinger, Inc.) for setting up simulations of biological and chemical systems, and is compatible with Visual Molecular Dynamics (VMD) for trajectory viewing and analysis.

Firefly (Computer Program)

Firefly, formerly named PC GAMESS, is an ab initio computational chemistry program for Intel-compatible x86, x86-64 processors based on GAMESS (US) sources. However, it has been mostly rewritten (60-70% of the code), especially in platform-specific parts (memory allocation, disk input/output, network), mathematic functions (e.g., matrix operations), and quantum chemistry methods (such as Hartree–Fock method, Møller–Plesset perturbation theory, and density functional theory). Thus, it is significantly faster than the original GAMESS. The main maintainer of the program is Alex Granovsky. Since October 2008, the project is no longer associated with GAMESS (US) and the Firefly rename occurred. Until October 17, 2009, both names could be used, but thereafter, the package should be referred to as Firefly exclusively.

On December 4, 2009, the support of any PC GAMESS versions earlier than the first PC GAMESS Firefly version 7.1.C was abandoned, and any and all licenses to use the code were revoked. Thus, users of the outdated PC GAMESS binaries (version 7.1.B and all earlier releases) were required to discontinue using the PC GAMESS and upgrade to Firefly.

On July 25, 2012, a state of the art edition of Firefly, version 8.0.0 RC, was launched for public beta testing. A relative comparison has shown that it is far faster and more reliable than the prior edition, Firefly 7.1.G. Many changes were made to enhance its abilities.

Materials Studio

Materials Studio is software for simulating and modeling materials. It is developed and distributed by BIOVIA (formerly Accelrys), a firm specializing in research software for computational chemistry, bioinformatics, cheminformatics, molecular dynamics simulation, and quantum mechanics.

This software is used in advanced research of various materials, such as polymers, carbon nanotubes, catalysts, metals, ceramics, and so on, by universities (e.g., North Dakota State University), research centers, and high tech companies.

Materials Studio is a client–server model software package with Microsoft Windows-based PC clients and Windows and Linux-based servers running on PCs, Linux IA-64 workstations (including Silicon Graphics (SGI) Altix) and HP XC clusters.

Software Components

- Analytical and Crystallization: to investigate, predict, and modify crystal structure and crystal growth.

  - Morphology

  - Polymorph Predictor

  - Reflex, Reflex Plus, Reflex QPA: to assist the interpretation of diffraction data for determination of crystallic structure, to validate the results of experiment and computation.

  - X-Cell: indexing for medium- to high-quality powder diffraction data from X-ray, neutron, and electron radiation sources.

- Quantum and Catalysis

  - Adsorption Locator: to find the most stable adsorption sites for various materials, including zeolites, carbon nanotubes, silica gel, and activated carbon.

  - CASTEP: to predict electronic, optical, and structural properties.

  - ONETEP: to perform linear-scaling density functional theory simulations.

  - DMol3: quantum mechanical methods to predict materials properties.

  - Sorption: to predict fundamental properties, such as sorption isotherms (or loading curves) and Henry's constants.

  - VAMP: high-speed calculation of a variety of physical and chemical molecular properties, e.g., for quick screening during drug discovery.

  - QSAR, QSAR Plus: to identify compounds with optimal physicochemical properties.

- Polymers and Classical Simulation: to construct and characterize models of isolated chains or bulk polymers and predict their properties.

- Materials Component Collection

- Materials Visualizer

Basic Workflow

- Materials Visualizer is used to construct/import graphical models of materials

- Accurate structure is determined by quantum mechanical, semi-empirical, or classical simulation

- Various required properties may be predicted/analyzed

MOPAC

MOPAC is a popular computer program used in computational chemistry. It is designed to implement semi-empirical quantum chemistry algorithms, and it runs on Windows, Mac, and Linux.

MOPAC2016 is the current version. MOPAC2016 is able to perform calculations on small molecules and enzymes using PM7, PM6, PM3, AM1, MNDO, and RM1. The Sparkle model (for lanthanide chemistry) is also available. This program is available in Windows, Linux, and Macintosh. Academic users can use this program for free, whereas government and commercial users must purchase the software.

MOPAC was largely written by Michael Dewar's research group at the University of Texas at Austin. Its name is derived from *Molecular Orbital PACkage*, and it is also a pun on the Mopac Expressway that runs around Austin.

MOPAC2007 included the new Sparkle/AM1, Sparkle/PM3, RM1 and PM6 models, with an increased emphasis on solid state capabilities. However, it does not have yet MINDO/3, PM5, analytical derivatives, the Tomasi solvation model and intersystem crossing. MOPAC2007 was followed by the release of MOPAC2009 in 2008 which presents many improved features.

The latest versions are no longer public domain software as were the earlier versions such as MOPAC6 and MOPAC7. However, there are recent efforts to keep MOPAC7 working as open source software. An open source version of MOPAC7 for Linux is also available. The author of MOPAC, James Stewart, released in 2006 a public domain version of MOPAC7 entirely written in Fortran 90 called MOPAC7.1.

Water Model

In computational chemistry, a water model is used to simulate and thermodynamically calculate water clusters, liquid water, and aqueous solutions with explicit solvent. The models are determined from quantum mechanics, molecular mechanics, experimental results, and these combinations. To imitate a specific nature of molecules, many types of models have been developed. In general, these can be classified by following three points; (i) the number of interaction points called *site*, (ii) whether the model is rigid or flexible, (iii) whether the model includes polarization effects.

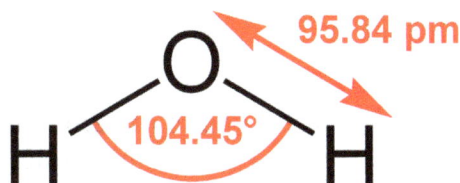

A water model is defined by its geometry, together with other parameters such as the atomic charges and Lennard-Jones parameters.

An alternative to the explicit water models is to use an implicit solvation model, also termed a continuum model, an example of which would be the COSMO Solvation Model or the Polarizable continuum model (PCM) or a hybrid solvation model.

Simple Water Models

The rigid models are considered the simplest water models which rely on non-bonded interactions. In these models, bonding interactions are implicitly treated by holonomic constraints. The electrostatic interaction is modeled using Coulomb's law and the dispersion and repulsion forces using the Lennard-Jones potential. The potential for models such as TIP3P (transferable intermolecular potential with 3-points) and TIP4P is represented by

$$E_{ab} = \sum_{i}^{\text{on } a} \sum_{j}^{\text{on } b} \frac{k_C q_i q_j}{r_{ij}} + \frac{A}{r_{OO}^{12}} - \frac{B}{r_{OO}^{6}}$$

where k_C, the electrostatic constant, has a value of 332.1 Å·kcal/mol in the units commonly used in molecular modeling; q_i and q_j are the partial charges relative to the charge of the electron; r_{ij} is the distance between two atoms or charged sites; and A and B are the Lennard-Jones parameters. The charged sites may be on the atoms or on dummy sites (such as lone pairs). In most water models, the Lennard-Jones term applies only to the interaction between the oxygen atoms.

The figure below shows the general shape of the 3- to 6-site water models. The exact geometric parameters (the OH distance and the HOH angle) vary depending on the model.

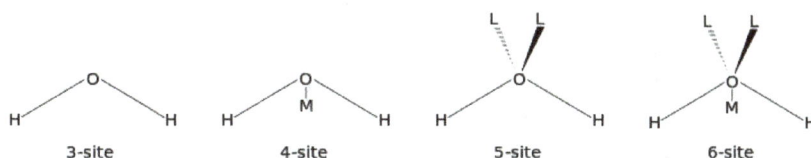

2-site

A 2-site model of water based on the familiar three-site SPC model has been shown to predict the dielectric properties of water using site-renormalized molecular fluid theory.

3-site

Three-site models have three interaction points corresponding to the three atoms of the water molecule. Each site has a point charge, and the site corresponding to the oxygen atom also has the Lennard-Jones parameters. Since 3-site models achieve a high computational efficiency, these are widely used for many applications of molecular dynamics simulations. Most of models use a rigid geometry matching that of actual water molecules. An exception is the SPC model, which assumes an ideal tetrahedral shape (HOH angle of 109.47°) instead of the observed angle of 104.5°.

The table below lists the parameters for some 3-site models.

	TIPS	SPC	TIP3P	SPC/E
r(OH), Å	0.9572	1.0	0.9572	1.0
HOH, deg	104.52	109.47	104.52	109.47
A × 10⁻³, kcal Å¹²/mol	580.0	629.4	582.0	629.4
B, kcal Å⁶/mol	525.0	625.5	595.0	625.5
q(O)	−0.80	−0.82	−0.834	−0.8476
q(H)	+0.40	+0.41	+0.417	+0.4238

The SPC/E model adds an average polarization correction to the potential energy function:

$$E_{pol} = \frac{1}{2} \sum_i \frac{(\mu - \mu^0)^2}{\alpha_i}$$

where μ is the dipole of the effectively polarized water molecule (2.35 D for the SPC/E model), μ^0 is the dipole moment of an isolated water molecule (1.85 D from experiment), and α_i is an isotropic polarizability constant, with a value of 1.608×10^{-40} F m². Since the charges in the model are constant, this correction just results in adding 1.25 kcal/mol (5.22 kJ/mol) to the total energy. The SPC/E model results in a better density and diffusion constant than the SPC model.

The TIP3P model implemented in the CHARMM force field is a slightly modified version of the original. The difference lies in the Lennard-Jones parameters: unlike TIP3P, the CHARMM version of the model places Lennard-Jones parameters on the hydrogen atoms too, in addition to the one on oxygen. The charges are not modified. Three-site model (TIP3P) has better performance in calculating specific heats.

Flexible SPC Water Model

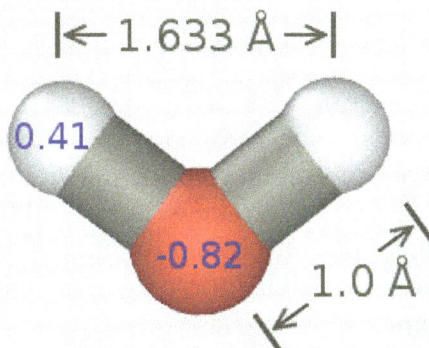

Flexible SPC water model

The flexible simple point charge water model (or Flexible SPC water model) is a re-parametrization of the three-site SPC water model. The *SPC* model is rigid, whilst the *flexible SPC* model is flexible. In the model of Toukan and Rahman, the O-H stretching is made anharmonic and thus the dynamical behavior is well described. This is one of the most accurate three-center water models without taking into account the polarization. In molecular dynamics simulations it gives the correct density and dielectric permittivity of water.

Flexible SPC is implemented in the programs MDynaMix and Abalone.

Other Models

- Ferguson (flex. SPC)

- CVFF (flex.)

- MG (flexible and dissociative)MG model

- KKY potential (flexible model). Molecular Simulation, 1994, Vol. 12(3-6), pp. 177–186.

- BLXL (smear charged potential). *J. Chem. Phys.* 110 (1999) 4566-4581.

4-site

The four-site models have four interaction points by adding one dummy atom near of the oxygen along the bisector of the HOH angle of the three-site models. The dummy atom only has a negative charge. This model improves the electrostatic distribution around the water molecule. The first model to use this approach was the Bernal-Fowler model published in 1933, which may also be the earliest water model. However, the BF model doesn't reproduce well the bulk properties of water, such as density and heat of vaporization, and is thus of historical interest only. This is a consequence of the parameterization method; newer models, developed after modern computers became available, were parameterized by running Metropolis Monte Carlo or molecular dynamics simulations and adjusting the parameters until the bulk properties are reproduced well enough.

The TIP4P model, first published in 1983, is widely implemented in computational chemistry software packages and often used for the simulation of biomolecular systems. There have been subsequent reparameterizations of the TIP4P model for specific uses: the TIP4P-Ew model, for use with Ewald summation methods; the TIP4P/Ice, for simulation of solid water ice; and TIP4P/2005, a general parameterization for simulating the entire phase diagram of condensed water.

Most of four-site water models use OH distance and HOH angle matching that of the water molecule in gas. An exception is OPC model on which no geometry constraints are imposed other than the fundamental C_{2v} molecular symmetry of the water molecule. Instead, the point charges and their positions are optimized to best describe the electrostatics of the water molecule. OPC reproduces a comprehensive set of bulk properties more accurately than commonly used rigid n-site water models. OPC model is implemented in AMBER force field.

	BF	TIPS2	TIP4P	TIP4P-Ew	TIP4P/Ice	TIP4P/2005	OPC	TIP4P-D
r(OH), Å	0.96	0.9572	0.9572	0.9572	0.9572	0.9572	0.8724	0.9572
HOH, deg	105.7	104.52	104.52	104.52	104.52	104.52	103.6	104.52
r(OM), Å	0.15	0.15	0.15	0.125	0.1577	0.1546	0.1594	0.1546
A × 10⁻³, kcal Å¹²/mol	560.4	695.0	600.0	656.1	857.9	731.3	865.1	904.7
B, kcal Å⁶/mol	837.0	600.0	610.0	653.5	850.5	736.0	858.1	900.0
q(M)	−0.98	−1.07	−1.04	−1.04844	−1.1794	−1.1128	−1.3582	−1.16
q(H)	+0.49	+0.535	+0.52	+0.52422	+0.5897	+0.5564	+0.6791	+0.58

Others:

- TIP4PF (flexible)

5-site

The 5-site models place the negative charge on dummy atoms (labeled **L**) representing the lone pairs of the oxygen atom, with a tetrahedral-like geometry. An early model of these types was the BNS model of Ben-Naim and Stillinger, proposed in 1971, soon succeeded by the ST2 model of Stillinger and Rahman in 1974. Mainly due to their higher computational cost, five-site models were not developed much until 2000, when the TIP5P model of Mahoney and Jorgensen was published. When compared with earlier models, the TIP5P model results in improvements in the geometry for the water dimer, a more "tetrahedral" water structure that better reproduces the experimental radial distribution functions from neutron diffraction, and the temperature of maximum density of water. The TIP5P-E model is a reparameterization of TIP5P for use with Ewald sums.

	BNS	ST2	TIP5P	TIP5P-E
r(OH), Å	1.0	1.0	0.9572	0.9572
HOH, deg	109.47	109.47	104.52	104.52
r(OL), Å	1.0	0.8	0.70	0.70
LOL, deg	109.47	109.47	109.47	109.47
$A \times 10^{-3}$, kcal Å^{12}/mol	77.4	238.7	544.5	554.3
B, kcal Å^6/mol	153.8	268.9	590.3	628.2
q(L)	−0.19562	−0.2357	−0.241	−0.241
q(H)	+0.19562	+0.2357	+0.241	+0.241
R_L, Å	2.0379	2.0160		
R_U, Å	3.1877	3.1287		

Note, however, that the BNS and ST2 models do not use Coulomb's law directly for the electrostatic terms, but a modified version that is scaled down at short distances by multiplying it by the switching function $S(r)$:

$$S(r_{ij}) = \begin{cases} 0, & \text{if } r_{ij} \leq R_L \\ \dfrac{(r_{ij} - R_L)^2 (3R_U - R_L - 2r_{ij})}{(R_U - R_L)^2}, & \text{if } R_L \leq r_{ij} \leq R_U \\ 1, & \text{if } R_U \leq r_{ij} \end{cases}$$

Thus, the R_L and R_U parameters only apply to BNS and ST2.

6-site

A 6-site model that combines all the sites of the 4- and 5-site models was developed by Nada and van der Eerden. Originally designed to study water/ice systems, however has a very high melting temperature

Other

- The effect of explicit solute model on solute behavior in biomolecular simulations has been also extensively studied. It was shown that explicit water models affected the specific solvation and dynamics of unfolded peptides while the conformational behavior and flexibility of folded peptides remained intact.

- MB model. A more abstract model resembling the Mercedes-Benz logo that reproduces some features of water in two-dimensional systems. It is not used as such for simulations of "real" (i.e., three-dimensional) systems, but it is useful for qualitative studies and for educational purposes.

- Coarse-grained models. One- and two-site models of water have also been developed. In coarse grain models, each site can represent several water molecules.

- Many-Body models. Water models built using training sets configurations solved quantum mechanically which then use machine learning protocols to extract potential energy surfaces. These potential energy surfaces are fed into MD simulations for an unprecedented degree of accuracy in computing physical properties of condensed phase systems.

Computational Cost

The computational cost of a water simulation increases with the number of interaction sites in the water model. The CPU time is approximately proportional to the number of interatomic distances that need to be computed. For the 3-site model, 9 distances are required for each pair of water molecules (every atom of one molecule against every atom of the other molecule, or 3×3). For the 4-site model, 10 distances are required (every charged site with every charged site, plus the O-O interaction, or $3 \times 3 + 1$). For the 5-site model, 17 distances are required ($4 \times 4 + 1$). Finally, for the 6-site model, 26 distances are required ($5 \times 5 + 1$).

When using rigid water models in molecular dynamics, there is an additional cost associated with keeping the structure constrained, using constraint algorithms (although with bond lengths constrained it is often possible to increase the time step).

References

- James J. P. Stewart (1989). "Optimization of parameters for semiempirical methods I. Method". Journal of Computational Chemistry. 10 (2): 209–220. doi:10.1002/jcc.540100208

- Hehre, Warren J.; Leo Radom; Paul v.R. Schleyer; John A. Pople (1986). Ab initio molecular orbital theory. John Wiley & Sons. ISBN 0-471-81241-2

- C. Møller; M. S. Plesset (1934). "Note on an Approximation Treatment form Many-Electron Systems". Physical Review. 46 (7): 618–622. Bibcode:1934PhRv...46..618M. doi:10.1103/PhysRev.46.618

- "Publisher's note: Sir John A. Pople, 1925-2004". Journal of Computational Chemistry. 25 (9): fmv–vii. 2004. PMID 15116364. doi:10.1002/jcc.20049

- Leach, Andrew R. (2001). Molecular modelling: principles and applications. Englewood Cliffs, N.J: Prentice Hall. ISBN 0-582-38210-6

- Allen, Frank (2002). "The Cambridge Structural Database: a quarter of a million crystal structures and rising". Acta Crystallogr. B. 58: 380–388. doi:10.1107/S0108768102003890

- James J. P. Stewart (1989). "Optimization of parameters for semiempirical methods II. Applications". The Journal of Computational Chemistry. Wiley InterScience. 10 (2): 221–264. doi:10.1002/jcc.540100209

- Silverstein, Robert M.; Francis X. Webster; David J. Kiemle (2005). Spectroscopy Identification of Organic Compounds. John Wiley & Sons, Inc. pp. 254–263. ISBN 978-0-471-39362-7

- Kevin J. Bowers; Ross A. Lippert; Ron O. Dror; David E. Shaw (2010). "Improved Twiddle Access for Fast Fourier Transforms". IEEE Transactions on Signal Processing. IEEE. 58 (3): 1122–1130. doi:10.1109/TSP.2009.2035984

- Lado, F.; Lomba, E.; Lombardero, M. (1995). "Integral equation algorithm for fluids of fully anisotropic molecules". The Journal of Chemical Physics. 103 (1): 481. Bibcode:1995JChPh.103..481L. doi:10.1063/1.469615

- Keeler, James (2010). Understanding NMR Spectroscopy. John Wiley & Sons, Inc. pp. 209–215. ISBN 978-0-470-74608-0

- Stillinger, F.H.; Rahman, A. (1974). "Improved simulation of liquid water by molecular dynamics". J. Chem. Phys. 60: 1545–1557. doi:10.1063/1.1681229

- Izadi, S.; Anandakrishnan, R.; Onufriev, A. V. (2014). "Building Water Models: A Different Approach". The Journal of Physical Chemistry Letters. 5 (21): 3863–3871. doi:10.1021/jz501780a

- Medders, G.R.; Paesani, F. (2015). "Infrared and Raman Spectroscopy of Liquid Water through "First Principles" Many-Body Molecular Dynamics". J. Chem. Theory Comput. 11: 1145–1154. doi:10.1021/ct501131j

- Hehre, Warren J.; Alan Shusterman; Janet Nelson (1998). Molecular Modeling Workbook for Organic Chemistry. Wavefunction, Inc. ISBN 1-890661-06-6

- Hohenberg, Pierre; Walter Kohn (1964). "Inhomogeneous electron gas". Physical Review. 136 (3B): B864–B871. Bibcode:1964PhRv..136..864H. doi:10.1103/PhysRev.136.B864

Permissions

Index